Java Digital Signal Processing

Metrowerks is proud to offer free world-class technical support on all its commercial products. Metrowerks will not provide technical support for CodeWarrior Lite, however you can access the benefits of Metrowerks' technical support by upgrading to a commercial product for as little as US$79. For more information, contact Metrowerks at 1-800-377-5416, or via email at sales@metrowerks.com.

Java Digital Signal Processing

Douglas A. Lyon

Hayagriva V. Rao

M&T BOOKS

M&T Books
A Division of MIS:Press, Inc.
A Subsidiary of Henry Holt and Company, Inc.
115 West 18th Street
New York, New York 10011
http://www.mispress.com

Printed in the United States of America

First Edition—1998

Library of Congress Cataloging-in-Publication Data

Lyon, Douglas A.
Java digital signal processing / by Douglas A. Lyon, Hayagriva V.Rao
p. cm.
ISBN 1-55851-568-2
1. Java (Computer program language. 2. Signal processing--Digital techniques. I. Rao, Hayagriva V. IITitle.
QA76.73.J38L96 1997
621.382'2'02855133—dc21 97-34785
CIP

MIS:Press and M&T Books are available at special discounts for bulk purchases for sales promotions, premiums, and fundraising. Special editions or book excerpts can also be created to specification.

For details contact: Special Sales Director
MIS:Press and M&T Books
Subsidiaries of Henry Holt and Company, Inc.
115 West 18th Street
New York, New York 10011

10 9 8 7 6 5 4 3 2 1

Associate Publisher: *Paul Farrell*
Production Editor: *Gay Nichols*
Copy Edit Manager: *Karen Tongish*
Managing Editor: *Shari Chappell*
Editor: *Laura Lewin*
Copy Editor: *Betsy Hardinger*

Acknowledgments

Thanks go to Victor Silva for his contribution of the first version of the VS package and assistance with the first port of the J++ version of DiffCAD; to Naoki Chigai for his contribution of the Java2HTML applet; to Damir Vamoser for his assistance with screen shots of the J++ compilations; to Vinny Antezzo for his first version of the AudioFrame digital oscilloscope; and to many other students, who put up with my beta code and manuscript efforts. I hope that this did not hurt my student evaluations too much!

Thanks go to Yanto Suryono for giving his permission to distribute the Surfer package with this book. Thanks also go to Don Gilbert for giving his permission to distribute the GR2PICT package.

Thanks go to the readers of the early drafts, particularly Mary Ellen Buschman, Fran Grodzinsky, and Vallaru Rao.

This work was made possible, in part, by a grant from the National Science Foundation DUE-9451520 and by a grant from the Educational Foundation of America.

This work was also made possible by support from the DeWitt Brothers Tool Company.

Contents in Brief

Contents

Chapter 4 Futil Recipes for Feudal Times 169

Preface

This book is for anyone who is interested in algorithms and applications of signal processing using Java. In this age of multimedia, that includes *many* programmers. Signal processing stands at the intersection of the frontiers of mathematics, scientific computing, and computer science. The programmer who buys this book probably has some programming experience and some mathematical maturity. Even those who skip the math will be rewarded by the Java capabilities that this book offers.

This book is our attempt to make up for some of things that make signal processing in Java hard. It's hard because the standard Java API is missing some key features. As a result, programmers have difficulty getting started in basic signal processing tasks.

This book contributes the DiffCAD program with all source code, containing more than 180 Java files and classes used to make a single programming environment. We also include complete source code, including original class libraries for the following:

- Reading and writing PPM and GIF image files
- Reading and writing Sun AU audio files
- Displaying waveforms using a digital oscilloscope and spectrum analyzer
- Audio and image processing tasks (including 1-D and 2-D FFT)
- Batch file processing, including a Java, C, and C++ to HTML converter
- Color conversion and raster to line segment conversion software
- Saving window contents to object PICT files (not bitmaps)
- Performing geometric and far-field diffraction
- Performing geometric image processing

This book is a result of a series of lectures given by Dr. Douglas Lyon, Director of the Image Sequence Processing Laboratory, at the University of Bridgeport. The courses formed a graduate-level computer science sequence in Java programming. The lectures for the course are transcribed into HTML and appear on the book's CD-ROM.

Chapter 1

Java and Its Promise

This chapter introduces Java's good points, and its not so good points. Any overview of Java must discuss the language specification and the language programming environment. Collectively, these are called *Java technology* and can include hardware as well as software.

This chapter is divided into Five main sections.

The first section, "What Is Java and Where Did It Come From", introduces the Java technology. We show that the term *Java* has come to mean both the Java programming language and the technology needed to support it.

The second section, "The Big Idea: WEBOS," describes the state of the art. We give an overview of the picoJava technology, the core of the Java chips that Sun intends to release. We also describe what may happen when Java appliances become widely used.

In the third section, "Java: The Good, the Bad, and the Ugly," we tell it like it is. Java has some good features, but it also has problems. This section outlines both. Please keep in mind that we really like Java (no, really). Still, we do not pull any punches. Every programming language has problems, and Java is no different. You should put your best foot forward, but watch where you are putting it.

The fourth section, "The HTML Model vs. the Java Model," describes the HTML (Hypertext Markup Language) model that is the basis of the World Wide Web and discusses the current problems with the diverse nature of data representations. You will also find diccourse on the big idea behind the Java model and how it may help to reduce the decoding problems that our Web browsers currently face.

In the fifth section, "The Java Developer Environments," we review a few software products—some of which are very useful. You're advised to select a programming tool with care. Often this means budgeting for more than one compiler and testing it yourself.

CD-ROM

The fifth section, "The Java Developer Environments," summarizes how you can get started using the software that comes with this book.

What Is Java and Where Did It Come From?

The term *Java* refers both to a language and to a technology that supports the language. When we speak of the *Java programming language*, we are talking about an object-oriented language developed by Sun Microsystems. Syntactically, this language is similar to several others. It has the braces of C, C++, and Objective C. It has the exception model of ZetaLisp, a flavors-based LISP that ran on LISP machines. The same method of exception handling has been proposed by Bjarne Stroustrup for use with the ANSI C++ standard [Spuler].

When we speak of *Java technology*, we are talking about the Java programming language and its support systems. These systems include a large library of classes called the *Java class libraries*. Java technology also includes a specification for run-time behavior, which is achieved using a Java machine specification. Java technology suggests that the Java machine may be implemented using any combination of software or hardware. When the Java machine is implemented in software, it is called the *Java virtual machine*.

Java started life as a language called Oak. It was designed to incorporate the best features of several languages into a single, new language. Just as important, features that were thought to make the language less reliable were left out. In balancing speed and reliability, Java's designers chose reliability. This design criterion makes Java inherently different from, for example, C++. In C++, features have been added to the language, but none have been removed. Further, it was important that C++ run as fast as C [Stroustrup].

One of Java's design objectives is that it be useful for distributed computing. In the distributed computing model, code can be downloaded for execution on demand in a secure fashion. Security became an important issue. If the code's source was not trustworthy, the code had to be treated as potentially harmful to users' data and computer hardware. Danger to users' data could include access to and distribution of sensitive information, such as credit card numbers, bank account numbers, and other proprietary data.

The Java machine specification is a Java language support technology that has become an integral part of the language. The tight integration of the Java language with the Java machine specification is one of Java's main selling points. The Java virtual machine places a layer of isolation between the running Java program and the underlying hardware, providing security and portability. The Java virtual machine provides security by optionally creating a security manager, that can keep a program from performing tasks that are considered a security risk. The Java virtual machine provides portability in that the virtual machine itself can run on several hardware platforms.

The Big Idea: WEBOS

If we consider all the Web servers on the Internet as being part of a large computer system, the Web is the largest operating system in the world. In fact, the Web's programming language is Java, and so, from this point of view, Java is an operating systems programming language. Sun plans to release Java machines that are not virtual. This means that the Java machines will be implemented in hardware. The operating system for these machines will be written in Java. People will no longer have to write cryptic C code to modify the Kernel of an operating system; they will be able to write in Java. When this happens, Java will probably spread into embedded system design until every appliance on the planet supports Java—even our toasters!

Consider the telephone. When unplugged from the network, the telephone is a useless piece of plastic, not worth the $20 it costs. The value added by the telephone is the network into which it is plugged. The same may be said of embedded systems on the Internet. For example, a toaster on the Net can download operational parameters (when to turn on, for how long, etc.), and it can use the network to communicate issues regarding its state (sorry to interrupt your net surfing, but the toast is done!).

Java chips will greatly reduce the price of an embedded Java controller. Dedicated chips will give embedded controllers speed and price advantages over their nonspecialized hardware counterparts. Sun is targeting the consumer market with mass sales of cheap chips.

Some targeted devices include TV set–top boxes, cellular telephones, pagers, digital TVs, smart VCRs, PDAs, printers, copiers, etc. In short, anywhere we find an embedded computer, Sun wants that computer to run Java. Because these chips will run byte code natively, there will be no need for a just-in-time compiler. Devices may not have a display, much memory, or any network connection. As a result, the application programming interface (API) for such embedded controllers will use a minimal API called the Java Embedded API. As of this writing, there is no published standard for the Java Embedded API. See http://www.sun.com for more information.

Sun Microelectronics, the Sun semiconductor division, calls its first chip architecture Java One. Sun plans to release two families of chips: microJava and ultraJava. MicroJava is a low-cost (less than $25) chip intended to target the embedded controller market. UltraJava is a higher-cost (less than $100) chip intended to target the workstation market. At the heart of the technology is a *super-scalar* stack-based reduced instruction set computer (RISC) machine called picoJava. It is super-scalar because it implements a four-stage pipeline that enables different parts of the processor to work on four different tasks at once. It is RISC because it executes most instructions in a single clock cycle.

Computing in a super-scalar pipeline is like using an assembly line. Data is passed from one worker to the next, and a process is applied to it. Figure 1.1 shows a sketch of the pipeline that, when filled, will permit picoJava to fetch, decode, execute, and cache, and then write back its results [Varhol].

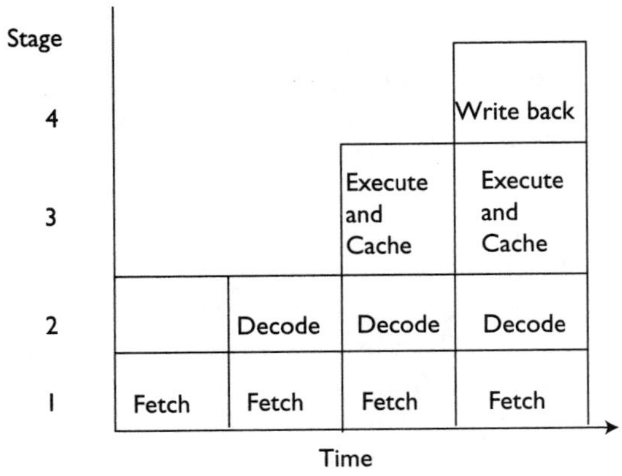

FIGURE 1.1 FOUR-STAGE PICOJAVA PIPELINE.

During the fetch operation, picoJava will load a 4-byte cache line into its processing stack. The stack consists of 64 on-chip 32-bit registers. After the on-chip storage is exceeded, RAM is used to implement the stack.

In addition to using ultraJava to target the workstation market, Sun hopes to use it to penetrate the network computer market. A network computer is a stand-alone computer connected to an enterprise's network infrastructure. The primary market consists of companies that want to centralize administration by maintaining a few servers. Such a scheme simplifies the deployment of applications by permitting them to be automatically downloaded over a network [Madany].

The proposed picoJava system shortens the path between the Java programs and the hardware by implementing the Java machine in hardware. This approach cuts out the *adapter layer* and uses Kona, a special operating system designed for the picoJava machine (Figure 1.2).

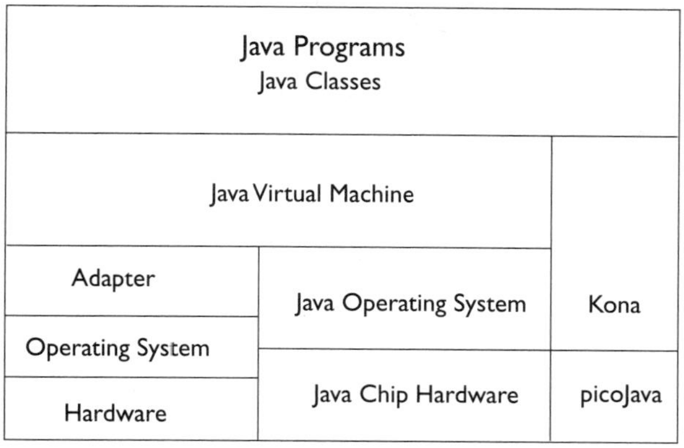

FIGURE 1.2 THE PICOJAVA KONA SYSTEM.

Keep in mind that picoJava is still in development. No silicon has been built, so no benchmarks have been run.

Java: The Good, the Bad, and the Ugly

Java is spreading through the computer science and engineering communities like wildfire, yet there is cause for caution. People are asking hard questions. Is Java suitable for engineering? Is Java suitable as a first programming language? Can Java be used throughout the computer science and engineering curricula? Is Java suitable for writing large programs? What are the drawbacks of being an early adopter of Java technology?

There is a lot of hype in the media today, and as a result, objective answers to these questions are not easy to come by. In fact, Java may not be suitable for writing large programs, and there may not be enough textbooks to use Java across the curriculum. Beta software is enough of a drawback to make any early adopter of a technology cringe. Being a beta tester of a compiler is not everyone's idea of a good time.

In this chapter, we attempt to balance our view of the language with a list of Java's good points, its bad points, and yes, its ugly points. We owe it to you, the reader, to say that being an early adopter of this technology comes at a cost: the time spent reading *many* Java books, writing custom libraries, buying new software, using beta compilers, and being the first (and sometimes only) Java programmer on the block. For us, the cost has well been worth it, but you must make your own decision. Use your judgment.

The Good

In this section we describe the good points of Java. Sometimes a good point about a language is also a bad point. For example, garbage collection is good, because it permits the programmer to forget about memory management during a programming task. Garbage collection simplifies design and eliminates a source of errors. Garbage collection is also bad, because it consumes system resources and could make Java unsuitable for low-level embedded control, a task for which it was intended.

Java Is a Strongly Typed Language

Java is a strongly typed language. All class names are treated as types and are used to check any reference to a class when passed as an argument to a method. Most modern languages have this feature, although the old-style C avoids it.

Java Is Small

Java is based on a small byte code interpreter. Including the self-contained microkernel, the byte code interpreter plus supporting classes is 215KB. This is a remarkable achievement. It means that byte code interpreters can reside on small ROMs and can give microcontrollers a means of running Java programs.

Java Is Portable

Java is a multi-platform language. Java's trademark is "Write Once, Run Anywhere." Because there is only one virtual machine specification, Java offers a standard, uniform programming interface to applets and applications on any hardware. The Java platform is therefore ideal for the Internet, where a program should be capable of running on any computer in the world. When you compile Java source, you obtain byte code. Byte code is output by the Java compiler and forms instructions to a Java virtual machine. Java is a portable language in that it can run on any hardware on which the Java virtual machine can run. Byte code is stored in a *class* file, that is downloaded to a Java virtual machine that contains a byte code interpreter. This concept is like the Pascal P-code of 20 years ago (which required a P-machine to execute the P-code) [Bowles]. When we speak of Java as a multiplatform language, we mean that it will run wherever there is an implementation of a Java virtual machine. Figure 1.3 shows the relationship between the Java program and the hardware.

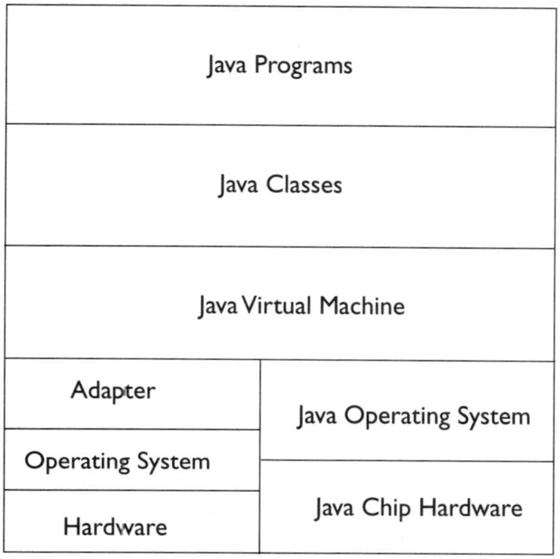

FIGURE 1.3 A SKETCH OF THE JAVA MODEL.

The multiplatform nature of Java is one of its strongest selling points. It can have a profound impact on how we evaluate computing resources. For the first time, we can benchmark precompiled code on a wide variety of platforms, ignoring compiler optimization for a specific machine. If a Java virtual machine is optimized for the hardware on which it runs, we can obtain a good measure of the machine's relative speed when running Java. Optimizing a Java virtual machine for specific hardware, however, is not an easy task. At present, for example, there are no Java virtual machines optimized for multiprocessor systems [Oaks et al.]. Thus, a threaded Java program cannot take advantage of the existence of more than one CPU. When this situation changes, Java will be a portable concurrent programming language.

Java Is Object-Oriented

Java is an object-oriented programming language. In an object-oriented paradigm, an instance of an object contains data as well as the algorithms needed to manipulate the data. In contrast, non–object-oriented programming languages pass data as arguments to procedures. Unlike Pascal, C, FORTRAN, or C++, Java has no functions. In Java, all *methods* reside in *classes*. C++ is a language with object-oriented extensions. This means that non–object-oriented programs can be written in C++, something that is generally not true in Java.

Java Has No Pointers

Java is more crash-proof than C, C++, and Pascal—a very good feature indeed. One reason is that Java does not provide a mechanism for manipulating pointers directly. Thus, there is no way for the programmer to obtain a memory address. Further, there is no pointer arithmetic and there are no pointer operations. Java eliminates the possibility of overwriting memory and corrupting data.

In C or C++ you can dereference a NULL pointer using

```
*ptr
```

When *ptr* is NULL, this code causes a segmentation fault error on UNIX or an immediate crash on some other machines. Sometimes pointers in C or C++ are pointing to illegal locations in memory. Accessing these locations can also cause a crash. This type of error is called a *dangling reference*. Another type of error, a *memory leak*, is created when data that has been discarded is not reclaimed with a *delete* statement. This can create an out-of-memory error that will crash the program (or computer) if it is not checked [Spuler].

There are many ways in which incorrect pointer use can crash a computer or program. We don't have enough space to list them all. We can be thankful that Java has no pointers.

Java Has No Multiple Inheritance

Multiple inheritance, as it was known in C++, has been eliminated in Java. Multiple inheritance is the ability to have two or more direct base classes. In 1966, Ole-Johan Dahl rejected multiple inheritance as a feature in Simula, reasoning that it would complicate garbage collection. Smalltalk does not support multiple inheritance either [Stroustrup 1994].

The problem with multiple inheritance is that duplicate class variables and method names must override each other, according to a policy that becomes a part of the language and can often be forgotten by the programmer. Elimination of multiple inheritance reduces the possibility that programmers will get confused about which method is in effect. It also limits the kinds of inheritance that can be performed. Java permits only an a-kind-of (AKO) taxonomy of class inheritance. Figure 1.4 shows an example of AKO class inheritance.

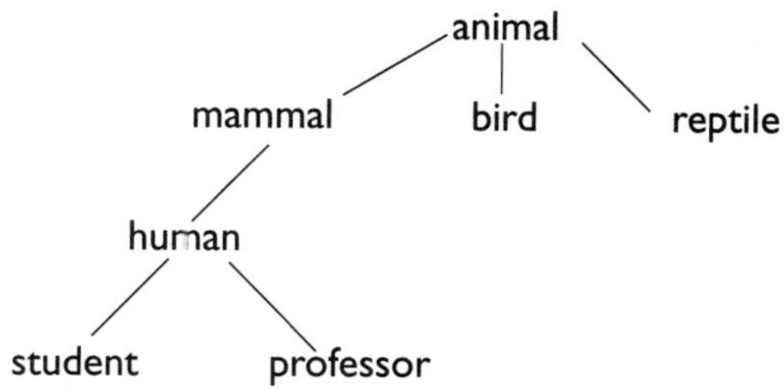

FIGURE 1.4 EXAMPLE OF AKO INHERITANCE.

In Java, an animal class can be extended to create a mammal subclass, thereby indicating that a mammal is a kind of animal. Because a student and a professor are both humans, they also inherit traits of mammals. In most other object-oriented languages, you can mix two or more AKO inheritance chains. Figure 1.5 shows an example of multiple inheritance.

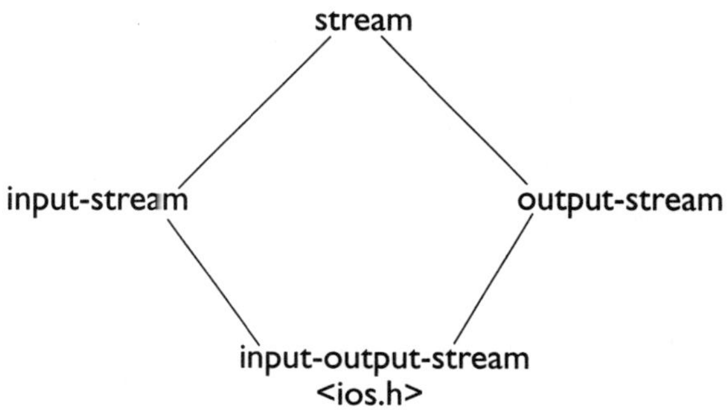

FIGURE 1.5 AN EXAMPLE OF MULTIPLE INHERITANCE.

With multiple inheritance, the attributes associated with the *stream* class, can be inherited by both the *input-stream* and the *output-stream*. When these two classes are joined by a third class, the *input-output-stream*, we have multiple inheritance. This language feature enables programmers to reuse the data structures and methods of two parent classes.

There is no doubt that method name ambiguity must be resolved at either compile time or at run time. And because Java must load libraries dynamically, it seems that the ambiguity would have to be resolved by the class loader. We suggest that one reason Java eliminated multiple inheritance was to promote implementation simplicity.

The removal of multiple inheritance is a design trade-off. The decision to remove multiple inheritance probably helped the code become more reliable, simplified garbage collection, and simplified the class loader. Programmers who are used to having multiple inheritance, however, will probably miss it.

Java Has No gotos

There are no gotos in Java. It is possible, however, to perform a multiloop break using the *break <label>* feature. Goto removal is a good thing, because it will probably lead to better codes and eliminate a common source of bugs.

Java Has No Global Variables

Java has no global variables. Instead, it gives you access to classes that have variables and methods. By restricting visibility, you can permit public access. When you apply access control to a class that has a static class variable, it could be argued that the variable is global to all other classes through the class name. The variable is accessed by *className. variableName*. For example, *Math.PI* is a global reference to a class variable.

Java Has No Macros

Java has no preprocessor and no macro language (as in RatFor, C, and C++). There are no compiler directives. It is not possible, as in C, to create language extensions. For example:

```
#define { Begin
```

This statement, which would create an ALGOL or Pascal-like style for code blocks, would confuse the code editor. It is not permitted in Java.

C/C++ macro facilities permit confusing function call errors. For example:

```
#define cube(x) x * x * x
x = cube(x+1);
```

This code expands into the following:

```
x = x + 1 * x + 1 * x + 1;
```

Thus, x becomes $3*x +1$ and not $x*x*x$.

Java Has Only Object Oriented Structures

There are no structures or union operations in Java. All the types are implemented as classes. This arrangement is probably a good thing, because the data manipulation methods can be built into the data structures.

Java Has Garbage Collection

Java has automatic storage operations, including a garbage collection mechanism. Garbage collection enables the Java virtual machine to reclaim storage used by discarded instances. You can explicitly invoke the garbage collector by using the *gc* method of the *System* class. The Java virtual machine will perform garbage collection without explicitly invoking *System.gc()*. The garbage collector is the only mechanism you can use to free storage in Java. To get an instance to be reclaimed by the garbage collector, you must remove all references to the object. Then either invoke *System.gc()* or wait for the garbage collector to reclaim the storage. The existence of the garbage collection mechanism in the Java virtual machine means that you will never have to worry about keeping track of storage.

Java Has Standard Class Libraries

Java has standard libraries that include an Abstract Window Toolkit (AWT). AWT enables object-oriented graphical user interface (GUI)–based programs to be portable. Previous attempts at this have not had much commercial success [Watson]. The class libraries have eight major packages, and the number is growing. An input/output package, *java.io*, enables a user to perform input and output stream manipulations. This is intended to make file and network data I/O manipulations into stream manipulations that work without direct involvement with the source of the stream. This approach provides a layer of abstraction that makes generic I/O programming much easier.

A network package, *java.net*, enables socket and uniform resource locator (URL) manipulations. The network package provides a standard, built-in method for turning sources and sinks of network data into streams. Once this occurs, the I/O package can be used to manipulate the streams.

A utilities package, *java.util*, contains several features considered standard in operating systems—features such as getting the date, time, random numbers, and so on.

These packages are a starting point for building portable programs. Any packages that build on these core Java packages will be portable. Also, the core Java packages are typically built into the systems that support Java. As a result, the core packages need not be downloaded every time a Java program needs something in a package. This arrangement makes the Java programs faster to download and makes the byte code files more compact.

Java Has boolean Types

Java uses a conditional statement that takes a boolean as an argument, whereas languages such as C or C++ permit an integer to be used as an argument to a conditional.

For example, in C or C++ it is possible to write the following:

```
if (1) {fprintf(stderr, "1 is not a boolean")};
```

In Java, the argument to a conditional must be of boolean type. Using an integer as an argument can create confusion between assignments and tests for equality. For example:

```
if (a = 0) {fprintf(stderr, "this will never be printed")};
```

This statement is a bug, because *a* will be assigned to the value 0. In fact, the == operator is needed:

```
if (a == 0) {fprintf(stderr, "this might be printed")};
```

With Java, the assignment-test confusion becomes a bug of the past.

Java Has Security

There is a class of programs called *applet viewers* that have their own Java virtual machines. Java-enabled browsers have built-in applet viewers.

Secure viewers protect the system by making an instance of the security manager. The security manager can disable operations that are considered dangerous (such as file I/O, creating consoles, or running native methods). Most Java-enabled browsers make an instance of the security manager.

Some applet viewers permit the running of Java programs without making an instance of the security manager. Thus, applets and applications are subject to the same security procedures.

A Java program that makes an instance of a frame, when an instance of the security manager is in place, will display an "untrusted Java Applet" label. This message alerts the user not to type sensitive data into the applet.

Java Has Exceptions

Java has a form of control structuring known as *exception handling*. Exception handling is a provision for handling abnormal circumstances that can prevent successful execution: subscript boundary violation, division by zero, overflow, I/O errors (from unavailable files or insufficient disk space), and so on. Java's exception handling mechanism can prevent a program from terminating abnormally. Exception handling is not a new idea and has been widely available in some languages (such as Ada, COBOL, C++, Delphi (http://www.borland.com), PL/I, and ZetaLisp) but not in others (such as BASIC, FORTRAN, Pascal, and C) [Goodenough]. Java's exception handling is closer to that of ZetaLisp than to that of any other language. Exceptions are subclassed in Java and so are treated in an object-oriented fashion (as opposed to their treatment in Ada, COBOL or PL/I).

e Java language specification identifies compile-time and run-time . For example, accessing an array index out of bounds is a run-time , according to the Java specification [Gosling et al.]. In C, the effect is endent on the operating system. For example, in Solaris (a Sun operating stem), this error causes a segmentation fault and a core file is dumped. On indows 3.x/95 or MacOS, the computer crashes. Windows NT may handle the error with a little more grace, but Java issues the following message:

```
java.lang.ArrayIndexOutOfBoundsException: 10
at TrivialApplet.test(TrivialApplet.java:18)
at TrivialApplet.main(TrivialApplet.java:12)
```

This alternative is civilized compared with crashing the computer.

Java Has Threading

Threading is a built-in feature of Java. A *thread* is a low-overhead context switch that enables a processor to change from one task to another very quickly. All the threads in Java could execute in parallel if the Java virtual machine existed that could take advantage of multiple processors. This is not the case, however, so only one thread can run at a given time (as of this writing). Threading is a high-level concurrent programming facility. In addition to Java, several other languages provide a high-level concurrent programming facility. Examples include Concurrent C, Concurrent C++, Concurrent Pascal, Concurrent Euclid, Modula-2, and Ada [Gehani].

In the past, multiple threads were programmed using support from the operating system. Java abstracts this relationship with the operating system by specifying how the virtual machine will behave when threading. There is nothing in the Java threads API that requires any operating-system involvement. In fact, the thread library of Solaris, on the Sun workstation, is unused in Java 1.0.

Java Has a Uniform Floating-Point Specification

Java uses the IEEE 754-1985 floating-point specification as a part of its language definition. Thus, round-off errors can be predicted in a platform-independent manner.

The Compilers Are Getting Fast

Java's compiler technology is improving. For example, just-in-time (JIT) compilers permit Java's "compile once, run anywhere" model to be as fast as compiled native code. The just-in-time compilers take the Java byte codes and compile them to native machine language. Our benchmark indicated an 18-fold increase in speed over interpreted byte codes.

This speed comes at a cost. Start-up time may be longer, although we could not verify this. Also, the JIT compiler is supposed to take more RAM, although we could not verify this, either. The overall increase in speed more than made up for any initial start-up costs on the DiffCAD program. Our benchmark was performed with a Metrowerks compiler running under MacOS. JIT compilers are not available on all platforms. If they were they would probably replace the interpreter model.

Strings Are First-Class Objects

Strings are not character arrays; rather they are instances of the *String* class. Thus, you must access them via method invocation and not as you access the array of characters in Pascal, C, or C++. This is a much cleaner way to manipulate strings and leads to better code.

Identifiers Have Unlimited Length

According to the Java language specification, identifiers may have an unlimited length. We have verified this for some very large values. For example:

```
      int
ThisIsAVeryLongNameInJavaWithMoreCharactersThanOneWouldTypicallyUseYouMayUseNumber
s1234567890ButNoOperators=0;
     int
ThisIsAVeryLongNameInJavaWithMoreCharactersThanOneWouldTypicallyUseYouMayUseNumber
s1234567890ButNoOperatorsAndTheyAreUnique=1;


System.out.println(ThisIsAVeryLongNameInJavaWithMoreCharactersThanOneWouldTypicall
yUseYouMayUseNumbers1234567890ButNoOperators+

ThisIsAVeryLongNameInJavaWithMoreCharactersThanOneWouldTypicallyUseYouMayUseNumber
s1234567890ButNoOperatorsAndTheyAreUnique
     );
```

The same principle should apply, in theory, to class names. Some development systems require that public classes be stored in files whose names match the class name. For these systems, it is not possible to have a public class identifier that exceeds file-name length limitations. These are implementation dependencies and not limitations imposed by the Java language specification.

The Bad

Sometimes Java's best features are also some of its worst features. For example, garbage collection has both good points (hence its listing in the preceding section) and bad points.

Sometimes Garbage Collection Is a Rotten Business

The drawbacks of garbage collection are as follows:

- The garbage collector can lead to nondeterministic program run times.
- For large systems garbage collection can use a significant amount of CPU time.

For example, during a time-critical interrupt, the Java virtual machine could sense that it is time for garbage collection, resulting in the loss of data, property, or even life. As far as we know, there is no way to turn off garbage collection from within the Java program. Some Java interpreters (such as the Metrowerks *javai*) have flags that disable asynchronous garbage collection. For example:

```
javai -noasyncgc
```

Keep in mind, however, that turning off asynchronous garbage collection doesn't mean you can stop worrying about it. The opposite is true. Turning off garbage collection means that you must invoke it yourself (or run out of memory).

If you have experience in programming large, garbage-collection–based systems (such as Lisp machines), you know that finding garbage is no easy task. The Lisp machine's *gc-immediate()* function, when run, started the garbage collection (just like Java's *System.gc()*).

Garbage collection in virtual memory typically causes a condition known as *thrashing*. Thrashing occurs when virtual memory is accessed nonsequentially. Thrashing causes different parts of the memory to be continually swapped in and out of disk. Keep in mind that RAM access time (measured in nanoseconds) is six orders of magnitude faster than disk access time (measured in milliseconds), so thrashing can cripple even the fastest of machines.

Java Is Not a Pure Object-Oriented Language

Java is not a *pure* object-oriented language. You cannot make an instance of any primative data type. The basic data types in Java are *boolean*, *int*, *long*, *float*, *double*, *char*, and *byte*. By contrast, in Smalltalk all data types are classes.

We Want Our Overloaded Operators!

Java does not permit the creation of overloaded operators. Contrast this with C++, which allows a programmer to give operators a context-dependent meaning. For example, in C++, the * operator can take two arrays as arguments and then multiply the arrays together. The Java designers did not appear to trust programmers to use the overloaded operator feature without writing cryptic code.

WARNING

In Java, the + operator is overloaded to concatenate strings. For example:

```
int x = 2; int y=3; String z = “4”;
System.out.println( x+z+y );
```

This code will treat *x, y,* and *z* as string objects and will output the following:

```
243
```

But the following code will treat *x* and *y* as numeric objects, add them, convert the result to a string, and concatenate the string with *z*.

```
int x = 2; int y=3; String z = “4”;
System.out.println( x+y+z );
```

Will output

```
54
```

Thus, the overloaded operators have become argument-dependent and have permitted the kind of cryptic code the Java designers wanted to avoid.

The API is Missing Many Features

Java is missing key features from its supporting API. For example, you cannot write to a serial port (even though almost every computer has one). You cannot (as of this writing) output to a printer from within a Java program. The Java language is new, and because features are missing from the API, Java may not be easy to use for some applications.

Java Has No Native Method Support for C++

You may want to extend the features of the Java API by programming in another language. Unfortunately the choice of language is currently limited to C.

There is no way, at present, to link between Java and C++, in part because of the problem of *name-space mangling*. In C++, source code function identifiers are mapped into different function names for the linker. This mapping is called name-space mangling. Functions are typically mangled according to their argument type. Different compilers may have different mangling schemes. Because Java has no way to know how functions will be mangled, the functions cannot be invoked.

The Java native language interface is not complete in JDK 1.0.2, but that is changing in JDK 1.1. When it is complete, circumstances are likely to get better.

The Ugly

No language is perfect, and Java has its design flaws. Some of them are harmless, if ugly. Others, such as the fragile base class problem, could limit Java's suitability for large software system development.

Arrays Can Be Allocated with Two Styles

Java supports the C and Java style of array allocation. In fact, the two styles are supported within the same statement. Thus, the following are three syntactically acceptable ways of specifying a two-dimensional array of *ints*:

```
int [][] i = new int[3][3];
int j[][] = new int[3][3];
// and now we put the Ug in Ugly!
int [] k [] = new int[3][3];
```

Java Has Fragile Base Classes

Java suffers from the fragile base class and the fragile interface problem. In Java, an interface can be used to store constants and to permit class and method specifications. For example, the DiffCAD program (used as a central example in this book) has an interface called *Constants* that contains a list of commonly held constants. Here is one line in *Constants*:

```
final double Pi_on_2 = Math.PI/2;
```

Suppose we add this line:

```
final double Pi_on_4 = Math.PI/4;
```

This code requires that every source code file that refers to *Constants* (in DiffCAD's case, seven files) to be recompiled. Including the linking phase, the recompilation takes 56 seconds on the authors machine, a PowerMac 8100/100 MHz with a PowerPC 601 and 72MB RAM. Another example is the abstract base class *Computation*. When *Computation* is altered, five files require recompilation; including the linking phase, 70 seconds elapse before the program begins to run. Thus, when programs become large, the fragile base class and fragile interface can seriously hamper programmers' productivity [Lewis].

"Appletcations" Are Confusing

The Java language has become a source of continuous confusion regarding the difference between an applet and an application. A package of classes in the core Java API is called the *java.applet* package, a very unfortunate naming convention. Within the *java.applet* package is a class called the *Applet* class. The *Applet* class is extended to create subclasses. Instances of *Applet* subclasses are called *Applets.*

Definition 1.1 An *applet instance* is an instance of a class that extends the Applet class.

Definition 1.2 An *application instance* is an instance of a class that contains a *main().*

Lemma 1.1 To run a Java application, it is necessary and sufficient both to have a *main()* and to invoke the *main().*

Corollary 1.1 Having a *main()* in a Java program is a necessary but not sufficient condition for running a Java application.

Lemma 1.2 To run a Java applet it is necessary and sufficient to extend the Applet class, implement the *init(),* and invoke the *init().*

Corollary 1.2 Subclassing the *Applet* class in a Java program is a necessary but not sufficient condition for running a Java applet.

NOTE

Corollary 1.1 follows directly from its parent, lemma 1.1. Similarly, corollary 1.2 follows from its parent, lemma 1.2. Definitions do not follow the construction of the lemmata.

In common use, the term *applet* has come to mean "a small Java application run from within a browser." We class such definitions as strictly incorrect. The following code is a segment of a large application, called DiffCAD, that dispatches a large number of different applets from within a Java application.

```
if (arg.equals("benchmark")) {
     AppletFrame w = new
          AppletFrame("BenchmarkApplet");
     String title ="BenchmarkApplet";
     String args[] ={""};
     w.startApplet("BenchmarkApplet",title,args);
}

if (arg.equals("surface")) {
     AppletFrame w = new AppletFrame("surface");
     String title ="surface";
     String args[] ={""};
     w.startApplet("surface",title,args);
}

if (arg.equals("search yahoo")) {
     AppletFrame w = new AppletFrame("Wa hoo!");
     String title "Search Yahoo";
     String args[] ={""};
     w.startApplet("SearchYahoo",title,args);
```

In fact, the applet is just a kind of *frame*. It runs in its own thread and has its own *applet context*. The point is that a large program can run many applets. A Java *application* is typically a program that contains a *main*. For example, the following is an application:

```
public class TrivialApplication {

     public static void main(String args[]) {
          System.out.println( "Hello World!" );
     }

}
```

An *applet* is an instance of an *Applet* subclass. For example:

```
import java.awt.*;
import java.applet.Applet;

public class TrivialApplet extends Applet {
     public void init() {
          repaint();
```

```
    }

    public void paint( Graphics g ) {
        g.drawString( "I am an Applet", 30, 30 );
    }

}
```

One popular Java reference states that a *class* becomes an applet by subclassing the *Applet* class and that an applet is an "embeddable window" [Chan and Lee]. No wonder even seasoned Java programmers misuse the term *applet.*

To top off the example, we present an appletcation that is an extension of the *Applet* class and contains a *main*:

```
import java.awt.*;
import java.applet.Applet;

public class TrivialApplet extends Applet{
    public void init() {
        repaint();
    }

    public static void main(String args[]) {
        System.out.println("An Appletcation");
    }

    public void paint( Graphics g ) {
        g.drawString( "Hello World!", 30, 30 );
    }

}
```

The preceding code may be called as an applet or as an application. When it is called as an applet, the `init()` method will be invoked and "Hello World" will be drawn. When it is called as an application, the *TrivialApplet* class is loaded, the *main* will be invoked, and the following will be displayed:

```
An Appletcation
```

TrivialApplet is a subclass of an *Applet* class but may be used as an applet or an application, depending on context. This permits the formulation of lemma 1.3 and corollaries 1.3a and 1.3b.

Lemma 1.3 Applets are run by invoking *init()*, and applications are run by invoking *main()*.

Corollary 1.3a The difference between an applet and an application is the invocation and not necessarily the content.

Corollary 1.3b Applets do not automatically run their *mains*. Applications do not automatically run their *inits*.

The Requirements to Match File Names and Class Names

Some compilers—such as JDK, J++, and the Symantec products—require that the file name of the Java source code match that of the public class name contained in the file. This requirement is not a part of the language specification [Gosling et al.]. It is a restriction imposed by the compiler implementation. This restriction is not uniformly imposed. For example, the Metrowerks CodeWarrior IDE for MacOS and Windows 95/NT do not impose this conformance between file names and class names.

This lack of uniformity makes it time-consuming to port source code from one compiler to another. We hope that Sun will correct this bug in its own compiler soon.

It is much easier to distribute one file with several small classes. The alternative is to make several small source files.

No Validation System

At present, no validation system exists for a Java compiler or supporting Java technology. This is a critical need, because many products appear to violate the compile-time and run-time specifications laid out by Sun.

A validation system would include series of test programs that would generate known compiler errors and known run-time errors. Such programs should elicit specific kinds of behavior from the Java class libraries.

The HTML Model vs. the Java Model

The HTML model permits a document to make references to files in other formats. A browser reads the references to the HTML files and dispatches them to a decoding program. For example, if the file is compressed, the browser may start a decompression program automatically.

The fatal flaw in this model is that browsers (and thei number of supporting applications, known as *helper apps*) can grow without bound. The Netscape browser, for example, recommends 16MB of RAM. As the applications become bloated, they tend to slow down even for simple tasks.

In this section we compare the HTML model with the Java model. In the Java model, code is compiled into class files and then downloaded, over the network, into an applet viewer. The applet viewer decodes the data stream that follows. The theory is that Java will become the language for decoding a wide range of data and that a browser need only support an applet viewer. For this model to work, the data must point to Java decoders that can be downloaded on demand.

The HTML Model

On the Internet, there are computers that run Hypertext Transfer Protocol (HTTP) servers. HTTP servers typically send data in response to a Web browser request. Generally, the data can be in any format; the HTTP server typically does not decode the data. As a result, Internet HTTP servers provide a wide variety of interesting and wonderful data formats to various browser-based clients. New formats appear all the time. Browsers typically understand some variant of HTML, and this arrangement has led to the formation of the HTML model.

In the HTML model, raw data is embedded in the HTML document by a hypertext reference (known as the *href* tag). To assist the browser in decoding the growing number of data formats, browsers use helper applications. In order to map the data to the correct helper application, browsers have a protocol that looks at the Multipurpose Internet Mail Extension (MIME) that the HTTP server transmits with the data. Based on the MIME, a lookup table determines how to decode and present the data. Figure 1.6 shows a screen capture of the presentation of one such table, known as the *helper window*.

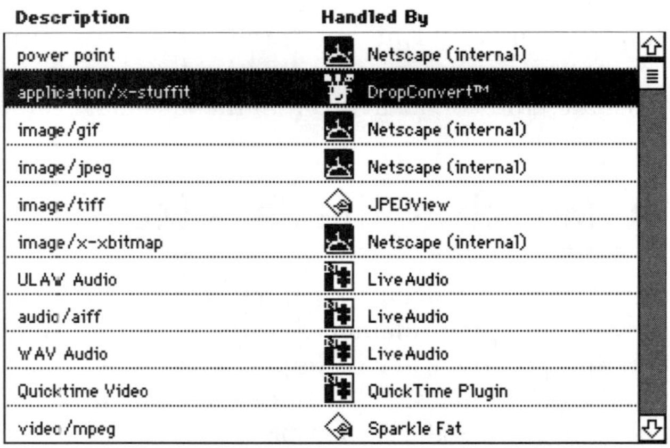

FIGURE 1.6 THE NETSCAPE HELPER WINDOW.

For each data type supplied by the HTTP server, there is a corresponding helper application, or *plug-in*. When this application is not present, the browser typically asks whether the user wants to save the file format. One of the authors has more than 77 items listed in the Netscape helper window. Naturally, these items do not represent all the possible data formats a browser can handle. You can customize a browser to handle any data format by launching a helper application. Thus, there are no limits to the number of data formats that may be present on the Web or handled by a browser.

Data will often be presented to the user in a variety of electronic forms, a veritable electronic tower of Babel. Suppose, for example, that a Microsoft Word document is to be supplied via the Web. You could supply it as a Word document, but Word 5 on a Mac cannot read Word 6 or 7 documents. So you could supply it as an RTF (rich text format) file so that Word 5 would understand most of it. Distributing Word documents using RTF to a variety of Word versions has some drawbacks: Some formatting will be lost, and some people will not have Word available as a viewer.

Word documents are often converted to HTML, which can be viewed by browsers the world over. Unfortunately, current versions of HTML can represent equations and vector graphics only as GIF images (a popular raster file format). Furthermore, HTML does not maintain the page layout of the original document. We could use PostScript, which enables users to download and print the document. Unfortunately, users may not be able to edit the document, and not all PostScript files will print to all printers. Adobe has stepped in with Portable Document Format (PDF), which lets you view the document on the screen and print it to all printers (if you have Adobe Acrobat). The problem, however, is that the user may still not be able to edit the PDF document.

The preceding example shows the rationale for having a wide variety of formats on the Web server. Having to have a different helper application for decoding each of these formats is cumbersome, and it's wasteful to store so many copies of the same content in different formats.

The Java Model

The Java model fixes some of the problems with the HTML model, but the Java model has yet to gain full acceptance.

In Java, compiled byte codes are stored in class files, which have a *.class* suffix. The class files are downloaded to the client's class loader. After a verification phase, the Java virtual machine (JVM) will interpret the byte codes. The role of the Java compiler is shown in Figure 1.7.

FIGURE 1.7 THE ROLE OF THE JAVA COMPILER.

Any browser with a JVM can load data decoders on demand. Imagine that you have a new image sequence compression scheme based on head-and-shoulders video. No one has your algorithm for decoding this new image format.

With Java, an algorithm for decoding a new data format can be downloaded on demand. This means that the Web has become object-oriented in the sense that the data can be joined with the program to manipulate the data. The Java model is a vast improvement over the current state of affairs, which requires that we have a wide variety of decoders on our hard drives. The role of the Java model on the network is shown in Figure 1.8.

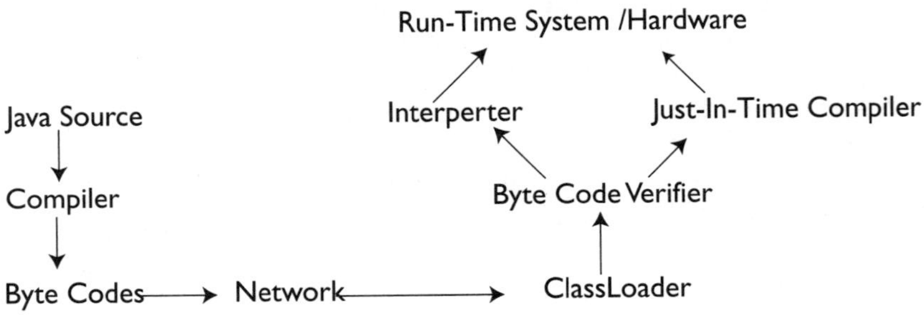

FIGURE 1.8 THE ROLE OF JAVA ON THE NETWORK.

The Java Developer Environments

Several alternatives are available for the development of Java, including Sun products (Java Workshop, Java Developers Kit (JDK)), Metrowerks' CodeWarrior, Symantec products (Visual Café, Visual Café Pro, Café), Natural Intelligence Roaster, Microsoft J++, Asymetrix SuperCede, and others. For more details, see http://www.javasoft.com/products/JDK/, ftp://ftp.metrowerks.com/pub/, http://www.metrowerks.com/, http://www.symantec.com, http://www.roaster.com/roaster/, and http://www.microsoft.com/java/.

J++ is available for Windows 95/NT. Workshop is available only for Solaris and Windows 95/NT. Roaster is available for MacOS. Symantec products and CodeWarrior are available for MacOS and Windows 95/NT. SuperCede is available for Windows 95/NT. JDK is available on all platforms (MacOS, Windows 95/NT, and Solaris). We will cover a few of these products in the following sections.

Feedback from students, and the authors' experience, has led to the conclusion that a good compiler is a worthwhile investment. Free tools (such as JDK) are good to have around, too. Some of the more expensive products are easier to use and faster to run. You should plan to purchase more than one compiler.

Some programmers would like to use code that has been written in C. As far as we know, the only compiler that supports linking to C is Metrowerks' CodeWarrior. Also, CodeWarrior comes with C, C++, Pascal and Java. CodeWarrior is an integrated development environment (IDE) that directly supports native method programming.

Getting Started on the Mac with CodeWarrior

CD-ROM

The CD-ROM that comes with this book has several example tools, and you're encouraged to try them. The primary example software is DiffCAD, a Java application that resides in several files. Upon opening the CD-ROM, the Mac user should look into the Mac folder to find a self-extracting archive such as the one shown in Figure 1.9.

FIGURE 1.9 THE DIFFCAD SELF-EXTRACTING ARCHIVE.

When you double-click on the **DiffCAD.sea** icon, you will be prompted with a standard file save dialog box, as shown in Figure 1.10.

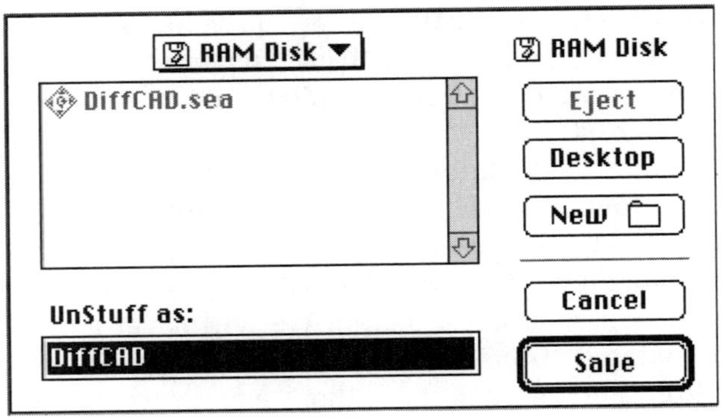

FIGURE 1.10 THE STANDARD FILE SAVE DIALOG BOX.

Once the self-extracting archive is expanded into the DiffCAD folder, double-click on the DiffCAD folder to find the CodeWarrior folder, shown in Figure 1.11.

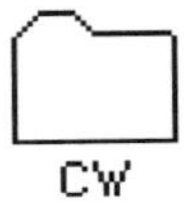

FIGURE 1.11 THE CODEWARRIOR FOLDER.

Within the CodeWarrior folder is a project that is prebuilt with CodeWarrior Professional Release 1 (the current version). If you double-click on the CodeWarrior folder, you will find a project file called **CW.11**, shown in Figure 1.12.

FIGURE 1.12 THE CODEWARRIOR PROJECT.

Double-click on the **CW.11** project file to start the CodeWarrior IDE. You will be presented with a project window, shown in Figure 1.13.

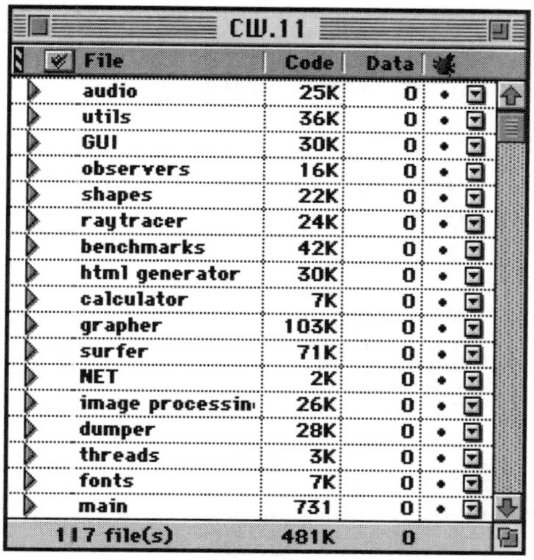

FIGURE 1.13 THE PROJECT WINDOW.

After the project window is displayed, you should be able to type **moth-r**. (On the Apple keyboard, the **moth** key has an logo as well as an icon that looks like a moth.) After the project begins to run, you will be presented with the display shown in Figure 1.14.

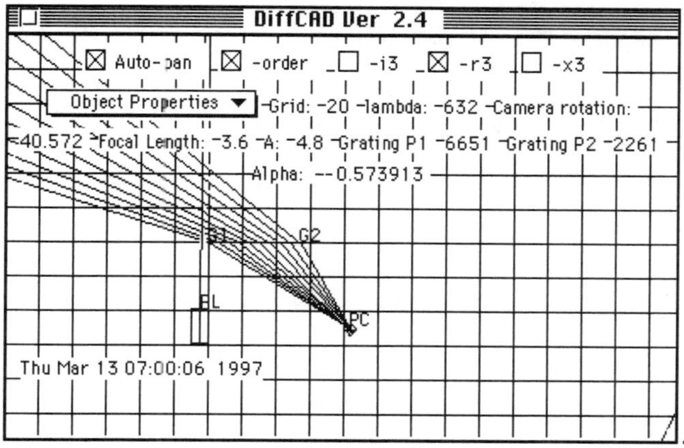

FIGURE 1.14 SCREEN SHOT OF THE DIFFCAD PROGRAM.

DiffCAD is a Java program that has been used to help design diffraction range finders.

Getting Started on Windows 95/NT with Metrowerks' CodeWarrior

As of this writing, Metrowerks has Professional Release 1 of the CodeWarrior IDE for the Windows 95/NT environment. It comes with Pascal, C, C++ and Java. There are also academic versions available at a discount.

CD-ROM

The project on the accompanying CD-ROM is in a file called **CW-WIN.ZIP**. When uncompressed, the file will place the DiffCAD folder on your hard drive. Double-click on the DiffCAD CodeWarrior project file, **DiffCAD.cwp**, and the project window will appear as shown in Figure 1.15.

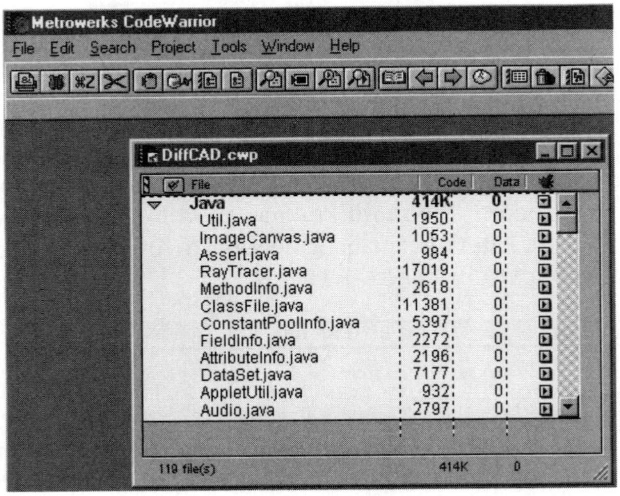

FIGURE 1.15 METROWERKS CODEWARRIOR DIFFCAD PROJECT ON WINDOWS 95.

Three strong points of CodeWarrior IDE are:

- CodeWarrior also compile Pascal, C, C++, and Java.
- The CodeWarrior IDE for Windows 95/NT has the same look and feel as the MacOS version.
- Metrowerks states that it intends to release a version of CodeWarrior for Solaris, a Sun UNIX operating system.

When the Metrowerks Unix release is ready, the CodeWarrior IDE will work on the Solaris, MacOS, and Windows 95/NT platforms.

Getting Started in Windows 95 with J++

In this section we give instructions for compiling the DiffCAD project on an Intel 486DX2 66MHz computer, running Windows NT 4.0 and Microsoft Visual J++ Professional Edition. We have also tested DiffCAD on a Windows 95 machine running on a 166 mmx equipped Pentium P5.

CD-ROM

Using an application called WinZip, open and uncompress the **DIFFCAD.ZIP** file on the CD-ROM as shown in Figure 1.16.

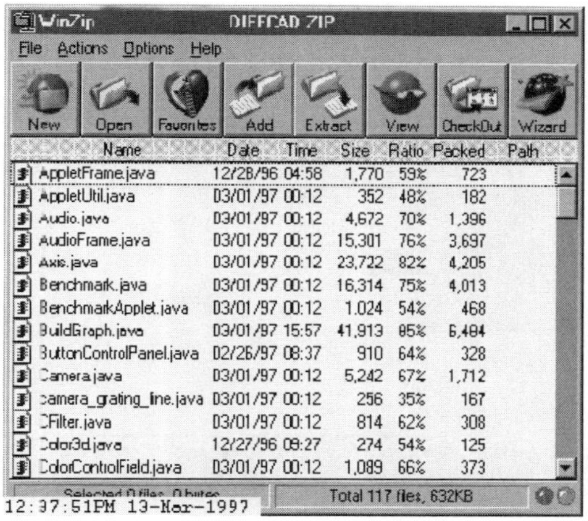

FIGURE 1.16 WINZIP 6.2 SHOWING A LIST OF FILES WITHIN DIFFCAD.ZIP.

WinZip was used to expand the files into the D:\work\DiffCAD directory. J++ provides a means to insert files into a project. All the files in the DiffCAD folder must be inserted into the J++ project.

Create a new project workspace of type **Project Workspace** in the DiffCAD directory as shown in Figure 1.17. From the Insert menu, select **Files into Project...** and add all the files in the DiffCAD directory to the project as shown in Figure 1.18.

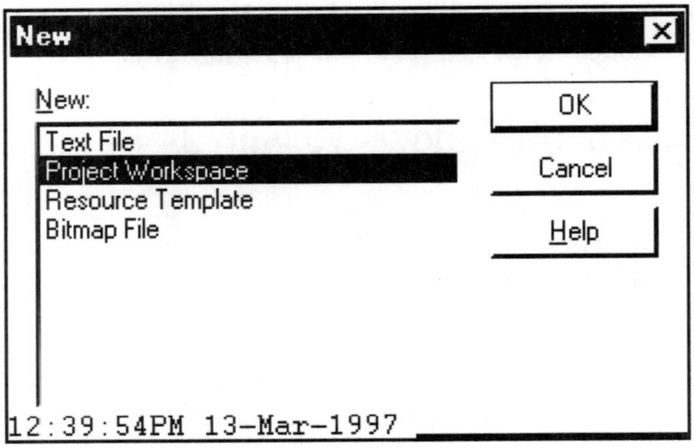

FIGURE 1.17 CREATING A NEW PROJECT WORKSPACE IN J++.

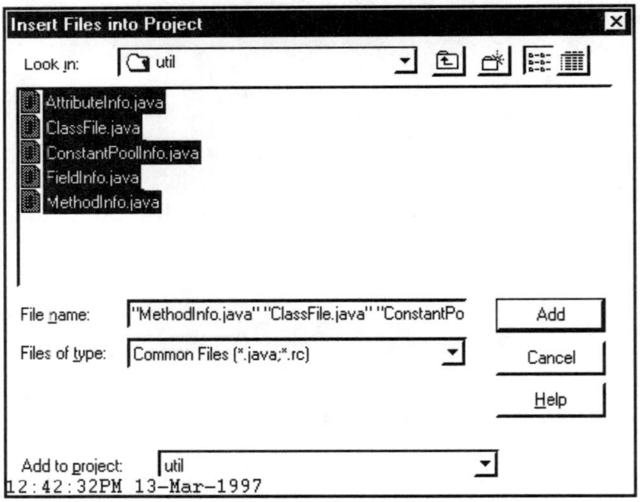

FIGURE 1.18 INSERTING THE FILES FROM THE UTIL PACKAGE INTO THE PROJECT.

This process takes several passes, because there are too many files to add in one pass (a limitation of the current. version of Developer Studio). Add the source files in the *Raytracer* and *util* directories to the project.

From the **Build ... Settings** menu selection, select the **Java** tab and set the warning level to **none**. Build and execute the project.

Two subprojects are assigned to the main DiffCAD project: *util* and *Raytracer*.

Figure 1.19 shows the *jview.exe* being invoked automatically by the J++ project manager.

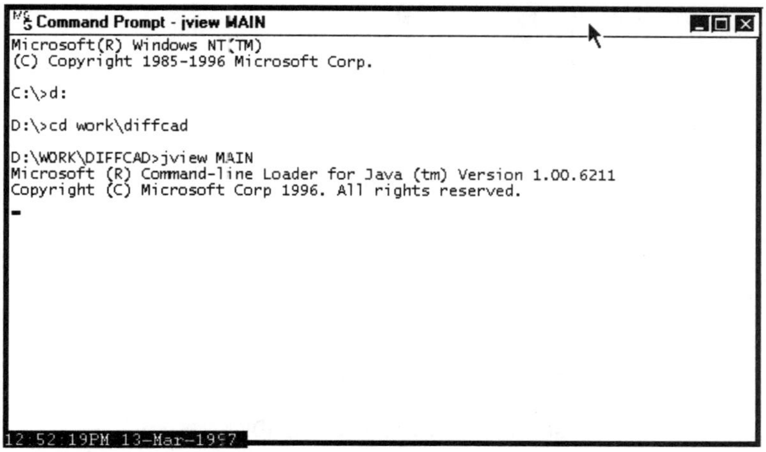

FIGURE 1.19 STARTING JVIEW.EXE ON THE MAIN CLASS.

Getting Started on the Mac with JDK 1.0.2

CD-ROM

In this section we describe the use of the Java Developers Kit, JDK 1.0.2, which is located on the CD-ROM. You are cautioned to try a different product because this one is freeware and is missing key features of some of the commercial products.

First, double-click on the **JDK.sea** icon and expand it to your hard drive. You will create a folder that looks like the one shown in Figure 1.20.

FIGURE 1.20 THE EXPANDED JDK ON DISK.

Inside the JDK folder you will find the Java compiler along with the four other items shown in Figure 1.21.

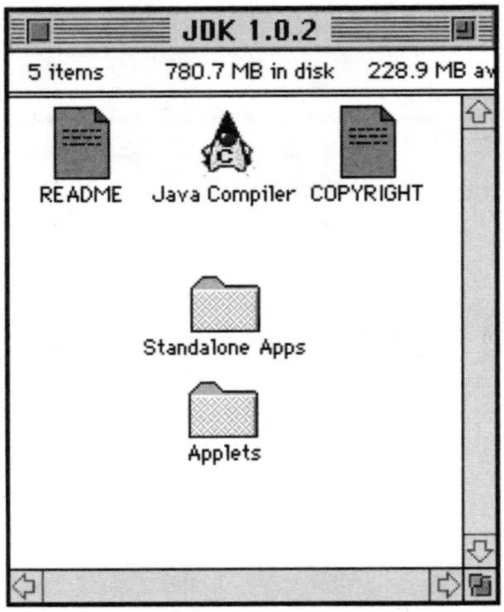

FIGURE 1.21 THE CONTENTS OF THE JDK FOLDER.

To make the Java compiler easier to use, make an alias to it and place it on your desktop. The Java compiler has a drag-and-drop interface, and having an alias will be of direct assistance in its use. Figure 1.22 shows how to find the alias command on the Mac.

FIGURE 1.22 FINDING THE ALIAS COMMAND ON THE MAC.

Once the Java compiler is on the desktop, uncompress the DiffCAD directory. An image of the DiffCAD self-extracting archive is shown in Figure 1.9.

Your first step is to compile the packages in the DiffCAD directory. There is no support for a UNIX-like Makefile command in the JDK version of the MacOS, so it must be done by hand. Find the two folders in the DiffCAD directory—Raytracer and util—shown in Figure 1.23.

FIGURE 1.23 THE TWO FOLDERS IN THE DIFFCAD DIRECTORY.

Double-click on the **Raytracer** folder and drag the Java source code to the Java compiler alias. Package compilation will create a Raytracer class folder near the **RayTracer.java** file, shown in Figure 1.24.

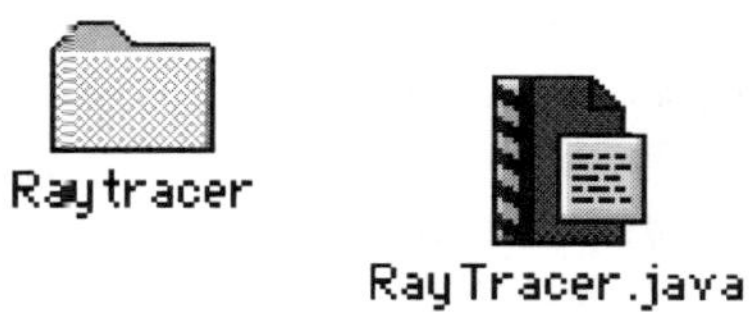

FIGURE 1.24 THE RAYTRACER CLASS FOLDER.

Move the class files from the RayTracer class folder into the parent folder (also called Raytracer). Repeat this process for the *util* package. Then, from the DiffCAD folder, select all the source files and drag them to the Java compiler alias. You may run out of memory during the compile. This is normal. Just view the files by name and drag the files with no corresponding class file, a few at a time, onto the Java compiler alias. It's time-consuming and error-prone. If you miss a file, your program will fail to load. After all the files are loaded, launch the Applet Viewer in the JDK 1.0.2/Applets directory. The Applet Viewer icon is shown in Figure 1.25.

FIGURE 1.25 THE APPLET VIEWER ICON.

Select the **file:properties** in the Applet Viewer and set the class access to **Unrestricted**, as shown in Figure 1.26.

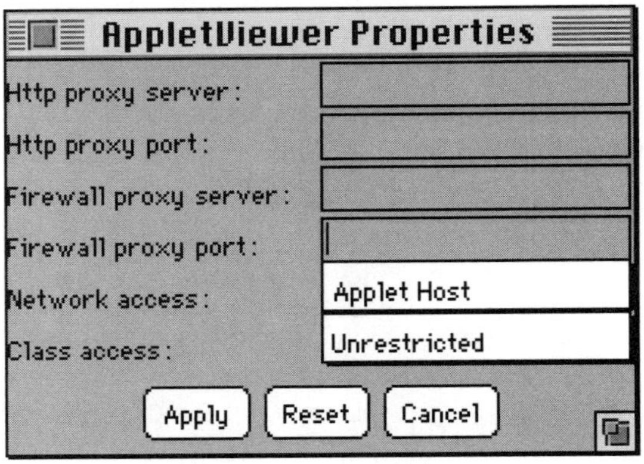

FIGURE 1.26 CLASS ACCESS MUST BE UNRESTRICTED.

The **Unrestricted** setting will enable DiffCAD to perform otherwise forbidden operations, such as opening a file and saving changes. The default class access is restricted, and it will cause a security manager exception to be thrown during normal DiffCAD usage. As of this writing, JDK 1.0.2 cannot turn off its restrictions on class access. This means that files cannot be opened with the MacOS version of the JDK.

Getting Started on the Mac with Symantec Café

In this section we describe a product called Symantec Café 1.5 for the PowerPC/MacOS. The IDE consistently emitted the following condition:

```
Error:  java.lang.OutOfMemoryError
```

This error occurs despite the allocation of 26MB to the IDE. The IDE supports the Sun Java compiler as well as the Symantec compiler. Switching to the Sun Java compiler within Symantec Café caused the program to hang.

Summary

In this chapter we discussed Java's good and bad points. Java's virtual machine is probably the reason Java is a successful technology. The Java language depends on a tightly integrated JVM and probably could not exist without it. Once the Java language and JVM were specified, flexibility depended on the Java class libraries. As soon as we transgress the boundary of the class libraries and JVM, Java becomes unsafe and nonportable. Java's growth depends on a growing set of portable class libraries. For those elements that are not portable (such as serial port support), the programming community must depend on Sun for API growth. This is quite a drawback.

This chapter also looked into the Java model as a means of creating decoders for Web browsers. Again, we noted that the Java class libraries lack the richness needed to support a wide variety of formats. Furthermore, Sun was cited for not advertising the API needed to manipulate an audio data stream. The Java language has so many demands that Sun cannot possibly support them all!

This chapter outlined Java's basic attributes: that it has no header files, macros, pointers, multiple inheritance, integer arguments to conditionals, structures, unions, or operator overloading. The language is portable and provides built-in garbage collection, a GUI library, array index checking, security features, threading, and exception handling. Java relies on the IEEE 754-1985 floating-point specification.

SHORTCUT

One of Java's big early drawbacks—that the byte code interpreter is slow—is being addressed. The JIT compilers blow away the byte code interpreters. Because of a smaller RAM footprint, byte code interpreters may be better for embedded systems. Even so, on desktop development systems the JIT compiler technology is a breakthrough.

The Java language confuses applets and applications and has mixed mode array declaration, confusing operator overloading for strings, fragile base classes, an inability to call C++, and an impoverished API. Java is not a pure object-oriented language. The API is generally improving, but the other drawbacks may be hard to fix.

No hype and no hyperbole: Java is a tool, and, like any tool, it is useful only for some applications.

Is Java suitable as a first programming language? As a teaching tool, Java is better than C or C++, but that damns Java with faint praise. Many first courses in programming are taught with C or C++, almost certainly because of industry demands. Now that industry appears willing to accept Java, introductory courses are switching to Java in great numbers.

As a tool for delivering cross-platform software, Java dominates any other technology we have seen. When we attempted to write portable programs in the past, we tried to use only the features supported by ANSI standards. Even under these circumstances, the porting of GUI-based code was tricky at best. DiffCAD, a program with over 180 Java files, ported to Java virtual machines running under MacOS, Solaris, and Windows 95/NT without recompilation.

Is Java suitable for engineering? That depends on the application. At this writing, Java has no support for serial ports. If the application involves data acquisition and processing via unsupported hardware (such as serial ports or video digitization cards), the answer could be no. Writing device drivers with interfaces to Java programs is not easy. Perhaps after native methods become easier to write, this situation will change.

Can Java be used throughout the computer science and engineering curriculum? Probably not yet. There are few college-level textbooks for Java that can target specific courses in computer science and engineering. We see a market for new books here.

Is Java suitable for writing large programs? The problem is Java's fragile base classes. Imagine if an include file, such as **<stdio.h>**, had to be changed. Makefiles across the UNIX operating system would have to recompile a large percentage of the code. Perhaps if developed libraries could be held stable during the course of development, Java could then be used to write large programs. Their maintenance, however, may prove impractical.

What are the drawbacks in being an early adopter of the Java technology? When we begin a trek into a new technological frontier, we must be willing to invest blood, sweat, and code. Programming is a humbling process. We have been writing code for many years, and programming in Java has meant recoding any previous work that we wanted to use. It has also meant having to deal with buggy beta compilers and spending many long and frustrating hours trying to get the answers to simple questions. It has meant being the first on the block to program in a new language. In some sense, this may isolate workers from one another. We have seen people take our Java courses and attempt to transfer the technology to their companies. Management can be slow on its feet, and some people are resistant to change. Perhaps the greatest drawback of being an early adopter of Java is the risk of being a boat-rocker. Go ahead and rock away!

In this chapter we compared the Java model with the HTML model. The basic question is, "Which will win?" As of this writing, the jury is out. Few, if any, decoders exist in pure Java, primarily because the Java API is impoverished. It can read only one audio format (monophonic, 8KHz, 8-bit, Sun AU files). This is a voice-grade audio file with about 48dB of signal-to-noise ratio (SNR) and a maximum frequency response of 4KHz. Compact disc recordings (CDs) are typically stereo 44.1 KHz with 16-bit samples (22.05 KHz cutoff with 96dB SNR). The Java API does not have a method for saving AU files. Furthermore, Sun does not advertise an API for manipulating the data stream that results from reading the AU files (something that this book addresses).

The Java API, as it currently stands, cannot process image sequences. The Sun API does not have a method for saving image files, another issue that this book addresses.

Sun has stated that there will be a multimedia API in the future but, as of this writing, has not said when that will happen. We have had to write a great deal of support code to make up for the areas where the API is currently lacking.

Chapter 2

Java Programming: The Basics

> "I don't know what the programming language of the year 2000 will look like, but I know that it will be named FORTRAN."
>
> —Tony Hoare, 1984

This chapter introduces you to Java. We assume that you are an experienced programmer in a structured language such as C or Pascal or that you have some background in object-oriented programming. We divide the subject of the Java language into two parts: syntax and semantics. We take a different path from that of the Java specification. The Java specification organizes the material by package, but we organize the material by concept. Thus, we place the wrapper classes (*Boolean*, *Character*, *Integer*, *Long*, *Float* and *Double*) into a single wrapper class subsection that resides in the section devoted to data types. We have found that the *java.lang* package does not present functions for maximum clarity of organization. For example, we prefer not to present threads and integers in the same subsection, they are basically unrelated topics even though they reside in the same package.

This chapter is divided into four main sections. The first section describes the notation we have devised to assist us with the presentation of Java syntax. We have devised this language to describe the grammar of Java. Because it is a language for describing languages, it is called a *metalanguage*. We base our metalanguage on a Backus Naur Form with some modifications. We refer to our version of the Backus Naur Form as Modified Backus Naur Form (MBNF).

Next, we discuss semantics of the Java language using examples. Often we will present several incarnations of Java usage in context. These incarnations are taken from working programs.

The second section covers the simple syntax of Java. Simple syntax includes those structures that are already familiar to C or C++ programmers. You may feel comfortable skipping or skimming this section.

In the third section we detail the data types in Java, including the object–oriented and the non–object–oriented data types.

The fourth section covers threads and provides several examples of multithreaded Java programs.

MBNF Notation

In this book we use MBNF to describe the syntax of Java. We use the translate functor of Prolog, ->, as the digraphic symbol meaning “can be written as” [Clocksin]. People who study formal languages refer to the -> symbol as a *production*. Java does not use the -> symbol, because Java has no pointers. We have found that the standard notation, as used in the Java specification, is hard to match on the blackboard during lectures. We wanted a compact notation whose typeset appearance would not deviate significantly from its hand-written appearance. This meant no bold or italic font could be used in the syntax definition.

The meta-symbols of BNF are shown in Table 2.1.

TABLE 2.1 BNF METASYMBOLS

Metasymbol	**Meaning**
->	can be written as
(X | Y)	grouping alternatively, X or Y
< >	syntactic construct, non-terminal symbol meta identifier
[X]	0 or 1 instance of X
{X}	0 or more instances of X
““	terminal
.	end of production

As an example, we show MBNF in MBNF:

```
syntax      ->  { production }.
production ->  identifier  "->"  expression "." .
expression ->  term { "|" term } .
term        ->  factor { factor } .
factor      ->  identifier |
                 quotedSymbol |
                 "("  expression  ")" |
```

```
                "["  expression  "]" |
                "{"  expression  "}" .
   identifier ->  letter { letter | digit } .
   quotedSymbol -> """ { anyCharacter } """ .
```

The syntax of a language will permit the formulation of a statement that compiles. The syntax does not describe the meaning of the statement however, nor does it describe the common usage. For this, we use examples and prose.

Following are the MBNF rules of Java which we will use in the remaining sections and chapters. These rules will, when used, always refer to their numbers. Sometimes the MBNF will be refined beyond the basic MBNF given here. When this occurs, it will not be preceded by the numeration.

```
compilationUnit  ->
          [packageStatement]    < importStatement >    < typeDeclaration > .

packageStatement  ->
          "package" packageName  ";"  .

importStatement  ->
          "import"  (  ( packageName  "."  "*"  ";"  )    |
          ( className  | interfaceName )  )  ";"  .

typeDeclaration  ->
          [docComment]  ( classDeclaration   | interfaceDeclaration )   ";"
          .

docComment  ->
          "/**"  "... text ..."  "*/"  .

classDeclaration  ->
          < modifier >  "class" identifier   [ "extends" className]
          [ "implements" interfaceName  <  "," interfaceName > ]   "{"  <
          fieldDeclaration >  "}"  .

interfaceDeclaration  ->
          < modifier >  "interface" identifier   [ "extends" interfaceName
          <  "," interfaceName > ]   "{"  < fieldDeclaration >  "}"  .

fieldDeclaration  ->
          (  [docComment]  ( methodDeclaration   | constructorDeclaration
          | variableDeclaration )  )   | staticInitializer   |  ";"  .

method_declaration  ->
          < modifier > type identifier   "("  [parameterList]  ")"  <  "["
          "]"  >   ( statementBlock  |  ";"  )  .

constructorDeclaration  ->
          < modifier > identifier  "("  [parameterList]  ")"
          statementBlock  .
```

```
statementBlock  ->
          "{"  < statement >  "}"  .

variableDeclaration  ->
          < modifier > type variableDeclarator
          <  "," variableDeclarator  >  ";"  .

variableDeclarator  ->
          identifier  <  "["  "]"  >  [ "=" variableInitializer]  .

variableInitializer  ->
          expression  |  (  "{"  [variableInitializer
          <  "," variableInitializer  >  [ "," ] ]  "}"  )  .

staticInitializer  ->
          "static" statementBlock  .

parameterList  ->
          parameter  <  "," parameter >  .

parameter  ->
          type identifier  <  "["  "]"  >  .

statement  ->
          variableDeclaration  |  ( expression  ";"  )  |  (
          statementBlock )  |  ( ifStatement )  |  ( doStatement )  |
          ( whileStatement )  |  ( for_statement )  |  ( tryStatement )
          |  ( switchStatement )  |  (  "synchronized"  "(" expression
          ")" statement  )  |  (  "return"  [expression]  ";"  )  |  (
          "throw" expression ";"  )  |  ( identifier  ":" statement )  |
          (  "break"  [identifier]  ";"  )  |  (  "continue"  [identifi-
          er]  ";"  )  |  (  ";"  )  .

ifStatement ->
          "if"  "(" expression  ")" statement  [ "else" statement]  .

doStatement ->
          "do" statement  "while"  "(" expression  ")"  ";"  .

whileStatement ->
          "while"  "(" expression  ")" statement  .

forStatement ->
          "for"  "("  ( variableDeclaration  |  ( expression  ";"  )  |
          ";"  )  [expression]  ";"  [expression]  ";"  ")" statement
          .

tryStatement ->
          "try" statement  <  "catch"  "(" parameter  ")" statement >  [
          "finally" statement]  .

switchStatement ->
```

```
        "switch"  "(" expression  ")"  "{"  <  (  "case" expression
        ":"  )  |  (  "default"  ":"  )  | statement >  "}"  .
expression  ->
        numericExpression  | testingExpression  |
        logicalExpression  | stringExpression  |
        bitExpression  | castingExpression  | creatingExpression  |
        literalExpression  |  "null"  |  "super"  |  "this"  |
        identifier  |  (  "(" expression  ")"  )  |  ( expression  (
        (  "("  [arglist]  ")"  )  |  (  "[" expression  "]"  )  |  (
        "." expression  )  |  (  "," expression  )  |  (  "instanceof"
        ( className | interfaceName )  )  )  )  .
numericExpression  ->
        (  (  "-"  |  "++"  |  "—"  )  expression )  |
        ( expression  (  "++"  |  "—"  )  )  |
        ( expression  (  "+"  |  "+="  |  "-"
        |  "-="  |  "*"  |  "*="  |  "/"  |  "/="  |  "%"  |
        "%="  )  expression  )  .
testingExpression  ->
        ( expression  (  ">"  |  "<"  |  ">="  |  "<="  |  "=="
        |  "!="  )  expression  )  .
logicalExpression  ->
        (  "!" expression )  |  ( expression  (  "&"  |  "&="  |
        "|"  |  "|="  |  "^"  |  "^="  |  (  "&&"  )  |  "||="  |
        "%"  |  "%="  )  expression  )  |  ( expression  "?" expres-
        sion  ":" expression )  |  "true"  |  "false"  .
stringExpression  =  ( expression  (  "+"  |  "+="  )  expression  )  .
bitExpression  ->
        (  "~" expression )  |  ( expression  (  ">>="  |  "<<"  |
        ">>"  |  ">>>"  )  expression  )  .
castingExpression  ->
        "(" type  ")" expression  .
creatingExpression  ->
        "new"  (  ( className  "("  [arglist]  ")"  )  |  ( type_speci-
        fier  [ "[" expression  "]" ]  <  "["  "]"  >  )  |  (  "("
        expression  ")"  )  )  .
literalExpression  ->
        integerLiteral  | floatLiteral  | string  | character  .
arglist  ->
        expression  <  "," expression  >  .
```

```
type  ->
          typeSpecifier  <  "["  "]"  >  .

typeSpecifier ->
          "boolean"  |  "byte"  |  "char"  |  "short"  |  "int"  |
          "float"  |  "long"  |  "double"  | className  |
          interfaceName  .

modifier ->
          "public"  |  "private"  |  "protected"  |  "static"  |
          "final"  |  "native"  |  "synchronized"  |  "abstract"  |
          "threadsafe"  |  "transient"  .

packageName ->
          identifier  |  ( packageName  "." identifier )  .

className  ->
          identifier  |  ( packageName  "." identifier )  .

interfaceName ->
          identifier  |  ( packageName  "." identifier )  .

integerLiteral ->
          (  (  "1..9"  <  "0..9"  >  )  |  <  "0..7"  > |
          (  "0"  "x"  "0..9a..f"  <  "0..9a..f"  >  )  ) [ "l" ]  .

floatLiteral  ->
          ( decimalDigits  "."  [decimalDigits]  [exponentPart]
          [floatTypeSuffix]  )  |  (  "." decimalDigits  [exponentPart]
          [floatTypeSuffix]  )  |  ( decimalDigits  [exponentPart]
          [floatTypeSuffix]  )  .

decimalDigits  ->
          "0..9"  <  "0..9"  >  .

exponentPart  ->
          "e"  [ "+"  |  "-" ] decimalDigits  .

floatTypeSuffix  ->
          "f"  |  "d"  .

character  ->
          "based on the unicode character set"  .

string  ->
          "\'"  < character  >  "\'"  .

identifier  ->
"a..z,$,_"  <  "a..z,$,_,0..9,unicode character over 00C0"  >  .
```

Simple Syntax

This section covers the aspects of Java that are familiar to an experienced programmer. Experienced C/C++ programmers will be tempted to skim or skip it. The differences between Java syntax and C/C++ syntax are small: documentation comments and an extension to the *for* loop. You will probably do fine if you read these two sections and then skip to data types.

NOTE

MBNF discussed in the previous section, will be reintroduced and sometimes extended as new syntactic features are covered.

Comments

There are three types of comments in Java:

- C–style comments that begin with /* and end with */
- C++–style comments that begin with // and end at the end of a line
- javadoc–style comments that start with /** and end with */

Comments in Java are just like those of C++. For example:

```
// This is a comment
/* and so is
     this
*/
```

There are good rules of style to be followed when using comments. For example, use C–style comments when you have multiple lines, because C++–style comments require the // in front of every line. For example:

```
/* This is a multi
     line comment
     it is easier to type
     because you don't need a
     // in front
     of every line.
*/
```

The C++–style comment is excellent for writing short comments. The end of the line denotes the end of the comment, and you don't have to worry about the terminating */.

Javadoc comments start with a /** and end with a */. The MBNF for the javadoc comment is as follows:

```
docComment  -> "/**"  "... text ..."  "*/"  .
```

CD-ROM

The JDK javadoc tool reads Java source code and scans for the javadoc comments. When they are encountered, javadoc generates HTML. These documentation comments are typically included before a class declaration, class member, or constructor. Code will work without them, but it is good practice to include them. They are not supported by all development tools and, more specifically, are not supported by Metrowerks CodeWarrior. As a result, we do not use the javadoc comments in this book. Instead, we have built a custom documentation generator that comes on the CD-ROM with this book.

Identifiers

An *identifier* starts with a letter and then contains letters or digits. It may not contain a keyword. Identifiers in Java may be of unlimited length. Recall rule 48:

```
identifier  ->

            "a..z,$,_"  <  "a..z,$,_,0..9,unicode character over 00C0"  >  .
```

The MBNF for the keyword follows:

```
keyword -> "abstract"  | "default" | "if" | "private" | "throw" | "boolean" | "do" |
"implements" | "protected" | "throws" | "break" | "double" | "import" | "public" |
"transient" | "byte" | "else" | "instanceof" | "return" | "try" | "case" | "extends" |
"int" | "short" | "void" | "catch" | "final" | "interface" | "static" | "volatile" |
"char" | "finally" | "long" | "super" | "while" | "class" | "float" | "native" |
"switch" | "const" | "for" | "new" | "synchronized" | "continue" | "goto" |
"package" | "this".
```

We were quite impressed when the following compiled and ran:

```
int
theUnlimitedLengthConstraintOfThe_Java_Identifiers_can_bring_1234Interesting_Varia
bles_into_your_code = 90;
```

We were also surprised when *goto* was listed as one of the keywords. Java does not support *goto* but reserves the *goto* keyword so that it may not be used as an identifier.

NOTE

Operators discussed next are not permitted as a part of the identifier. Identifiers are used in several places in Java, as discussed in the following sections.

Operators

Java has assignment operators like those of most other languages. There are also the shortcut operators of C and C++. Here is a typical assignment operator followed by the shortcut version common to C, C++, and Java:

```
i = i + 1;
i++;
```

Both versions have the same effect. The + operator is a binary operator, because it takes two arguments. The ++ operator is a unary operator, because it takes only one argument. The Unary operators are +, -, ++, ~,!.

Unary operators group right-to-left, so that the following two statements mean the same thing:

```
-!i
-(!i).
```

The MBNF for the Java operators follows:

```
Operator ->  "=" | ">" | "<" | "!" | "~" | "?" | ":" | "==" | "<=" | ">=" | "!=" |
        "&&" | "||" | "++" | "--" | "+" | "-" | "*" | "/" | "&" | "|" | "^" | "%" |
        "<<" | ">>" | ">>>" | "+=" | "-=" | "*=" | "/=" | "&=" | "|=" | "^=" |
        "%=" | "<<=".
```

In Java, you can have prefix unary operators:

```
++i;
```

You can also use postfix unary operators:

```
i++;
```

Unary operators work just like their C counterparts. The order of precedence is listed next, along with the MBNF.

```
postfix_operators->
            "[]" | "." | "(" <parameters> ")" | "++" | "--" .

unary_operators ->
             "+"| "-" | "++" | "--" | "~" | "!" .

creation_operators->
             "new" | "(" <type> ")" .

multiplicative_operators->
             "*" | "/" | "%".

additive_operators ->
             "+" | "-".

shift_operators ->
             "<<" | ">>" | ">>>".

relational_operators ->
             "<" | ">" | ">=" | " <=" | "instanceof" .

equality_operators ->
             "==" | "!=" .

bitwise_AND_operator  ->
             "&" .

bitwise_XOR ->
             "^" .

bitwise_OR ->
             "|" .

logical_AND ->
             "&&" .

logical_OR ->
             "||" .

conditional_operator ->
             "?:" .

assignment_operators ->
             "=" | "+=" | " -=" | "*=" | "/=" | "%=" | ">>=" | "<<=" |
            ">>>=" | "&="  | "^="  | "|=".
```

It is a part of the C idiom that assignment operators be augmented with other operators. We will give examples of the augmented operators, "+=" , "-=" , "*=" , "/=" , "%=" , ">>=" , "<<=" , ">>>=" , "&=" , "^=" and "|=" , as well as the condition operator, because they are the most commonly misused by our students.

Consider this example:

```
if (theCowsComeHome) {System.out.println("moo");}
     else System.out.println("no milk for you!");
```

It is the same as the following:

```
System.out.println(
     theCowsComeHome ? "moc" : "no milk for you!");
```

So the conditional operator requires a Boolean type expression followed by an expression to return if true with a :, and then an expression to return if false.

Here are some examples of the augmented operators with their non-augmented counterparts:

```
i *= 10; // is the same as:

i = i * 10;
i += 10; // is the same as:
i = i + 10;
i /= 10; // is the same as:
i = i / 10;
i >>= 3; // is the same as:
i = i >> 3; // a right shift with sign in the extension
i >>>= 9; // is the same as:

i = i >>> 9; // a right shift with zeros in the extension
```

The following code example will reverse the bits in an int. Bit reversal code is used to perform transforms such as the Fast Hartley Transform (FHT) and the Fast Fourier Transform (FFT):

```
int bitr(int j) {
     int ans = 0;
```

```
    for (int i = 0; i< nu; i++) {
        ans = (ans <<1) + (j&1);
        j = j>>1;
    }
    return ans; }
```

The variable *nu* is declared as an int and is equal to the number of bits to be processed. It is common to replace such bit shifting code with lookup tables that group the bits into 8– or 16–bit groups.

Flow of Control

This section describes Java's features for altering the flow of control. Java is strongly typed, so *if* , *for,* and *while* statements require Boolean expressions as arguments. In fact, all flow of control statements require Boolean expressions, with the single exception of the *switch* statement, which permits an integer type expression.

Expressions

Central to the use of the flow of control is the notion of a Boolean type expression. Unlike C and C++, Java will not accept an expression that is of integer type. The following MBNF defines the expression statement in Java along with the testing expression and the logical expression:

```
expression ->
numericExpression
| testingExpression   | logicalExpression   |
stringExpression  | bitExpression   |
castingExpression   | creatingExpression   |
literalExpression   |  "null"   |
"super"  |  "this"   |
identifier   |  (  "(" expression  ")"  )   |
( expression   (  (  "("  [arglist]  ")"  )   |
(  "[" expression  "]"  )   |  (  "." expression  )   |
(  "," expression  )   |
(  "instanceof"  ( className | interfaceName )  )   )  )  .
numericExpression  ->
(  (  "-"   |  "++"   |  "--"  )  expression )   |
```

```
( expression   (  "++"   |  "--"  )  )   |
( expression   (  "+"   |  "+="   |  "-"
|  "-="   |  "*"   |  "*="   |  "/"   |  "/="   |  "%"   |  "%="  )
expression  )  .
testingExpression  ->
( expression
(  ">"   |  "<"   |  ">="   |  "<="   |  "=="   |  "!="  )
expression  )  .
logicalExpression  ->
(  "!" expression )   |  ( expression
(  "&"   |  "&="   |  "|"   |  "|="   |  "^"   |  "^="   |
(  "&&"  )   |  "||="   |  "%"   |  "%="  )  expression  )   |
( expression  "?" expression  ":" expression )   |  "true"   |  "false"  .
```

In the following sections we give examples of expressions that alter the flow of control.

If

The *if* statement in Java requires a boolean type expression:

```
ifStatement ->
            "if"  "("  expression  ")" statement   [ "else" statement]  .
   if (type.compareTo("ConstantValue") == 0) {
   if (i >= out.length) {
   if (e.target == saveSound_mi) {
   if (e.target == graphSound_mi)  {
```

The *then* part of the *if* may take the shape of any valid statement. No {} characters are required:

```
if (Math.abs(return_val) < 0.99)
    return return_val;
else return 0;
```

An extended clause is often indented to indicate when the clause has ended. Sometimes it is good style to indicate the end of the clause with a comment. For example:

```
if (interfaces == null) dos.writeShort(0);
 else {
    dos.writeShort(interfaces.length);
```

```
            for (int i = 0; i < interfaces.length; i++)
                dos.writeShort(
                    ConstantPoolInfo.indexOf(
                        interfaces[i], constantPool));

} // end else
```

This practice is particularly good when the beginning of a clause is off the screen. Poor use of indentation and comments can make code hard to read. Consider the following example:

```
if(x1 < datarect.x) x1 = datarect.x;
      else if(x1 > datarect.x + datarect.width )
          x1 = datarect.x + datarect.width;


if(y1 < datarect.y) y1 = datarect.y;
       else if(y1 > datarect.y + datarect.height )
          y1 = datarect.y + datarect.height;
```

Now compare the preceding code with the following code:

```
  if (x1 < datarect.x) x1 = datarect.x;
  else if (x1 > datarect.x + datarect.width )
           x1 = datarect.x + datarect.width;


if (y1 < datarect.y) y1 = datarect.y;
 else if (y1 > datarect.y + datarect.height )
           y1 = datarect.y + datarect.height;
```

NOTE

Else if is used as a word pair, significantly cleaning up and shortening the code.

This example, is a long *if else* chain:

```
if(xminText.equals(e.target)) {
```

```
    xmaxText.requestFocus();
    return true;
  } else
  if(xmaxText.equals(e.target)) {
    yminText.requestFocus();
    return true;
  } else
  if(yminText.equals(e.target)) {
    ymaxText.requestFocus();
    return true;
  } else
  if(ymaxText.equals(e.target)) {
    xminText.requestFocus();
    return true;
}
```

Now let's convert it into the *else if* style:

```
if(xminText.equals(e.target)) {
     xmaxText.requestFocus();
     return true;
} else if(xmaxText.equals(e.target)) {
          yminText.requestFocus();
          return true;
} else if(yminText.equals(e.target)) {
          ymaxText.requestFocus();
          return true;
} else if(ymaxText.equals(e.target)) {
          xminText.requestFocus();
          return true;
}
```

Such long dispatches are common in Java.

While and Do Statements

The syntax of the *while* statement is given by the following MBNF:

```
whileStatement ->

   "while"  "(" expression  ")" statement  .
```

Like the *if* statement, the *while* statement must have a Boolean type expression. Also, the expression and ensuing statement will be evaluated as long as the Boolean expression returns true. If the expression returns false, the statement will not be executed. If the statement throws an exception, the *while* statement will also throw one. The *while* statement always evaluates the expression firs, before evaluating the statement. To change this order, use the *do* statement. The MBNF for the *do* statement is as follows:

```
doStatement ->

        "do" statement  "while"  "(" expression  ")"  ";"  .
```

The *do* statement will always evaluate the statement at least once before evaluating the *while* statement. The *do while* structure of Java is just like the *repeat until* structure of Pascal. It is a compile-time error to pass a non-Boolean type expression to the *do while.*

To make an infinite loop, just make the boolean expression a constant *true*:

```
while (true) {
               draw();
               try {Thread.sleep(1000);}
               catch (InterruptedException e) {}
          }
```

It is common to embed assignments in the boolean expression given as an argument to the *while* statement. For example:

```
while ((next = tokens.nextToken()) != tokens.TT_EOF) {
```

Note the required use of the parentheses. The following would result in a compilation error:

```
while (next = tokens.nextToken() != tokens.TT_EOF) { //BUG!
```

That's because the assignment operator takes lowest precedence.

The *tokens.nextToken()* sets the *next* int to a value and checks the value against the *tokens.TT_EOF* class variable, an action that returns a boolean.

Switch

The *switch* statement transfers control to one of several statements depending on the value of an expression. It is a compile-time error if the expression is not constant or one of char, byte, short, or int. The following is the MBNF for the *switch* statement:

```
 switchStatement ->
"switch"  "(" expression  ")"  "{"   <  (  "case" expression  ":"  )   |  (
"default"  ":"  )   | statement >   "}"  .
```

The use of the *switch* statement usually involves a *break*:

```
switch (expression) {
     case const_1:
          statement1;
          break;
     case const_2:
          statement2;
          break;
     default:
          statement3;
          break;
}
```

If you omit the *break*, any remaining branches are executed. You can use only one default branch in the *switch* statement. It is typical in Java to have long *switch* statements when dispatching keyboard events. For example:

```
public boolean keyDown(Event e, int key) {
     switch (key) {
             case 'o':
                  openAudioStream();
                  return true;
            case 'v':
               saveAs();
                  return true;
             case 'p':
                  play();
                  return true;
             case 'n':
                  normalize();
                  return true;
```

```
case 'm':
     am();
     return true;
case 'f':
     fm();
     return true;
case '^':
     sawWave();
     return true;
case 'e':
     ulawData=Audio.encodeUlaw(doubleData);
     return true;
case '1':
     fft();
     return true;
case '2':
     ifft();
     return true;
case '3':
     dft();
     return true;
case 't':
     timeDelay();
     return true;
case 'g':
     graphSound();
     return true;
case 'd':
     doubleData = Audio.decodeUlaw(ulawData);
     return true;
case 's':
     sineWave();
     return true;
case '[':
     squareWave();
     return true;
```

```
            case 'T':
                  triangleWave();
                  return true;
            case 'r':
                  ulawData = Audio.reverse(ulawData);
                  return true;
            case 'u':
                  graphUlaw();
                  return true;
      }
```

Keyboard shortcuts are popular with experienced users. It is unfortunate that the *switch* statement can take only scalar values. The same test is typically performed for menu-item instances, with lengthy code as the result.

For

The *for* statement is 10 times more common than the *while* statement in the DiffCAD program. The MBNF for the *for* statement follows:

```
forStatement ->
          "for"  "("  ( variableDeclaration  |  ( expression  ";"  )  |
          ";"  )   [expression]  ";"   [expression]  ";"   ")" statement
          .
variableDeclaration  ->
          < modifier > type variableDeclarator   <  "," variableDeclarator
          >  ";"  .
variableDeclarator  ->
          identifier  <  "["  "]"  >  [ "=" variableInitializer]  .
variableInitializer  ->
          expression   |  (  "{"  [variableInitializer   <  ","
          variableInitializer  >  [ "," ] ]  "}"  )  .
```

If the test is not a Boolean type expression, a compile-time error results. If the test is not satisfied, the statement will not execute. The statement need not execute, at all. Here are examples of the *for* statement:

```
for (i=1; i < 99; i++) {
     }; // null statement
for (;;) { // an infinite loop
     if (expression) { break}
```

```
        // more junk here
}
for (int i = 99; i < 50; i++)
        System.out.println("I never printed");
```

The comma (,) is permitted in the initialization and increment section of the *for* loop. For example:

```
for (i=0, j=10; i < 100; i++, j +=2) {
//more stuff here
}
```

If the initializations or update parts of the *for* loop throw an exception, the *for* loop will throw an exception. The *for* loop is the only statement in Java that uses the comma as a separator.

Continue

The *continue* statement in Java aborts an iteration. It is a compile-time error to have a *continue* in something other than a *while*, *do*, or *for* statement. A *continue* statement with no label identifier proceeds to the next enclosing iteration. A *continue* statement with a label identifier proceeds to the next enclosing *labeled* statement. The MBNF for the continue statement follows:

```
continue_statement ->

 "continue"  [identifier]  ";"
```

For example:

```
foo: for (int i = 1; i < 5; i ++) {
for (int j=1; j < 5; j++) {
     if ( i % j == 2) {
          System.cut.println("continue");
          continue foo;
     }
     System.out.print(i*j + " ");
  }
     System.out.println();
}
```

The preceding code outputs the following:

```
1 2 3 4
2 4 continue
3 6 9 12
4 8 12 16
```

Here's more typical usage:

```
for (int i = 1; i < constantPool.length; i++) {
          if (constantPool[i] == null)
               continue;
       // more stuff follows
   }
```

Here, *continue* will proceed to the next *i* without finishing the rest of the loop.

The labeled *continue* appears to be less popular than the unlabeled *continue*. DiffCAD does not have a single use of the labeled *continue*.

Break

The *break* statement appears much more often in DiffCAD than the *continue* statement (345 times vs. 15 times). Here's the MBNF for the *break* statement:

```
break_statement ->

"break"  [identifier]  ";"
```

It is a compile-time error not to enclose *break* within a break target. Valid break targets are labels, *switch*, *while*, *do*, or *for* statements. *Break* can be used with or without an identifier label. *Break* causes control to pass to the innermost enclosing break target. The break target completes normally. Compare this to a *continue* statement in a loop. *Continue* will continue with the loop; *break* will break out of it.

The following example permits the *for* loop to terminate normally for two reasons. The first is the *for* loop expression, *n < 2*Math.PI*, which is tested only at the top of the loop. The second is the *(i >= out.length)* expression, which is tested at the bottom of the loop.

```
done: for (double n = 0; n < 2*Math.PI; n = n + step) {
```

```
                out[i] =  in[i]* Math.sin(n);
                i++;
                if (i >= out.length) { break done;}
        }
```

In the following case, the *break* is removed by adding a more complex test in the *for* statement.

```
for (double n = 0;
                (n < 2*Math.PI) && (i < out.length);
                n += step, i++) {
                out[i] =  in[i]* Math.sin(n);
        }
```

Sometimes the *break* statement is essential, as in a *switch* statement.

Return

The *return* statement in Java takes an optional expression. It transfers control to the invoker. The MBNF for the return statement follows:

```
return_statement ->

"return"  [expression]  ";"
```

When the expression is omitted, *void* is returned.

Data Types

Data types in Java are divided into two basic categories: reference types and primitive types. The reference types consist of three sub types: class types, interface types, and array types. We defer discussion of interface types and array types until later. The following two sections describe the primitive types and class types.

Primitive Types

Java has several primitive types. Primitive types can store primitive values, each of which needs a specific number of bits of storage and has a specific precision.

The primitive data types of Java are sometimes also called *scalar* types. There are two kinds of scalar types: *boolean* and *numeric*. Boolean types may have two predefined values: *true* or *false*. Numeric types can be subclassed into two subtypes: *integer* or *floating point*. There are five kinds of integer types and two kinds of real types, float and double. Figure 2.1 shows the taxonomy of the primitive types in Java.

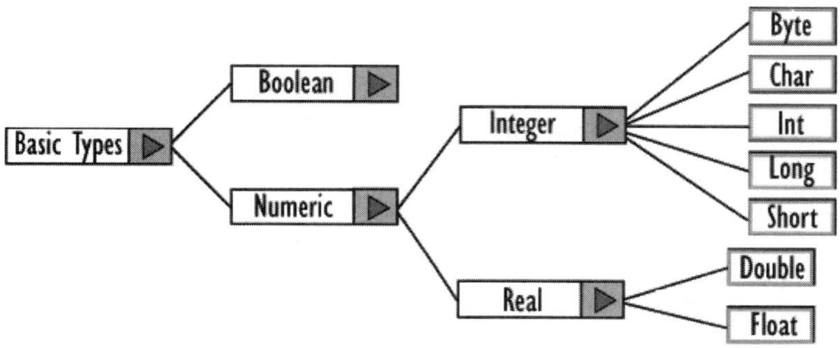

FIGURE 2.1 TAXONOMY OF PRIMITIVE TYPES IN JAVA.

Each of the primitive data types is assigned a specific amount of storage. Its value and range are shown in Figure 2.2.

type name	value	value	
boolean	1 bit	true, false	
char	16 bit	unicode	character
byte	8 bit	signed	$-2^{8-1}...2^{8-1}-1$
short	16 bit	signed	$-2^{16-1}...2^{16-1}-1$
int	32 bit	signed	$-2^{32-1}...2^{32-1}-1$
long	64 bit	signed	$-2^{64-1}...2^{64-1}-1$
float	32 bit	IEEE-754-1985	6-7 sig. figs.
double	64 bit	IEEE-754-1985	14-15 sig.figs.

FIGURE 2.2 VALUES AND RANGES FOR PRIMITIVE DATA TYPES.

All variables in Java are held as undefined until set.

For example, in the following code, a variable, *x* , was declared but not initialized before being accessed.

```
public class TrivialApplication {

    public static void main(String args[]) {
        int x;
```

```
        System.out.println( x );
    }
}
```


This is a compile–time error in Java, and the compiler emits the following:

```
Error    : Variable x may not have been initialized.
TrivialApplication.java line 7     System.out.println( x );
```

All integer data types (byte, short, int and long) are signed. All characters in Java are Unicode, an international standard [Unicode]. The character type is defined as a 16-bit unsigned unique code value. For example:

```
char c = 'a';
```

It is possible to assign a numeric literal to a character-typed variable and assign a character literal to an integer variable:

```
char theChar = 48;
integer theValue = 'a';
```

All reals are stored as single–precision floating–point numbers, called *float*, or double–precision floating–point numbers, called *double*. Java reals use the IEEE 754-1985 format [IEEE]. The smallest and largest positive nonzero values for floats range from 1.40239846ee-45 to 3.40282347e+38 (*Float.MIN_VALUE* and *Float.MAX_VALUE*). The smallest and largest positive nonzero values for doubles range from 4.94065645841246544e-324 to 1.79769313486231570e+308 (*Double.MIN_VALUE* and *Double.MAX_VALUE*).

Now for a reality check. Using the Metrowerks CodeWarrior we discover the floating bugs in the Java floating–point libraries:

```
class fp_error {
    public static void main(String argv[]) {
        float fmin = Float.MIN_VALUE;
        float fmax = Float.MAX_VALUE;
        double dmin = Double.MIN_VALUE;
        double dmax = Double.MAX_VALUE;
        String b = " ";
```

```
        // bug in FP
        System.out.println(fmin + b + fmax +
            b + dmin + b + dmax);
    }
}
```

The preceding code outputs the following:

```
1.4013e-45 3.40282e+38 2.22507e-308 1.79769e+308
```

Please note that round(1.40239846e-45) = 1.4024e-45 and not 1.4013e-45, as CodeWarrior indicated. It has output the following:

```
3.40282e+38 instead of 3.40282347e+38
2.22507e-308 instead of 4.94065645841246544e-324
1.79769e+308 instead of 1.79769313486231570e+308
```

Note that the 2.22507e-308 is off from 4.94065645841246544e-324 by 16 orders of magnitude. The problem with the output from *println* is also documented in the language specification (20.10.15). In short, it says that *println* renders finite values in the same form as the %g format of the printf function in the C programming language, which can lose information because it produces at most six digits after the decimal point." In fact, our search of the *println* source code shows that the implementation depends on a call to a *String* class conversion method called *toString*. This explains why the error exists in the *toString* method as well as in the *System.out.println*.

Finally, the wrong value for the *MIN_VALUE* variable is Sun's fault. If you look at the source for *Double.java* (for the Mac and Windows versions of the 1.0.2 API), it declares *MIN_VALUE* as 2.225 e-308 instead of 4.94e-324:

```
public static final double MIN_VALUE = 2.2250738585072014E-308;
```

We have verified this with cross–platform testing (Solaris, Windows 95/NT, and MacOS). The HTML docs, however, show the correct value (as do all books we have seen on Java). The version 1.1 source code may correct this problem.

The following program shows how to code the escape sequences into Java:

```
package stringUtilities;
public interface Char {
    char backspace = '\b';
    char horizontalTab = '\t';
    char newLine = '\n';
    char formFeed = '\f';
    char carrageReturn = '\r';
    char doubleQuote ='\"';
    char singleQuote ='\'';
    char backSlash = '\\';
    char maxOctal = '\377';
    char minOctal = '\000';
    char maxUnicode = '\uFFFF';
    char minUnicode = '\u0000';
}
```

Named Constants

Constants are values that remain unchanged during their life. For example:

```
static final double PI = 3.14159265358979323846;
```

The term *static* is known as a modifier. It indicates that there is to be only one incarnation of the field. *Final* indicates that *PI* cannot change during the life of the *PI* variable. Assignment of a *final* field is a compile-time error. You may, if you like, restrict the visibility of the *static final* field by using a modifier prefix. You can perform computations with knowns on the right-hand side of the equals sign. For example:

```
    private final double pi_2 = Math.PI * 2;
```

Here we see that the field *pi_2* has a *private* visibility. This means that the *pi_2* name is not visible outside the class (a discussion we defer until later). Any valid type specifier may follow the final modifier. The MBNF for this is as follows:

```
typeSpecifier ->
```

```
"boolean"   | "byte"   | "char"   | "short"   | "int"   | "float"   | "long"   |
"double"   | className   | interfaceName  .
```

The following are examples of the final modifier:

```
static final int  HORIZONTAL = 0;
static final String version = "1.0";
public static final int ACC_PUBLIC = 0x1;
static final float  twoPI =  (float) (2*Math.PI);
public static final char NOT_CODE = (char)12;
private static final long UNIT = 1000;
```

Note the use of 0x1 to denote hexadecimal code. Also note the use of the cast operators (*float*) and (*char*). Casting is run-time type conversion and is discussed in more detail later in the chapter.

Classes

This section introduces a reference type known as a *class*. A class is a combination of code (methods) and data (variables) joined into a single entity. A class is essentially a description of how to make an instance of an object. This is what makes Java object–oriented. Every instance has a class that is used to determine how to create the instance, which variables the instance will contain, and which messages the instance will respond to. Instances of classes are of reference type. In Java a reference is just like a pointer, except that no pointer arithmetic is permitted.

The following is a summary of the MBNF needed to declare a class:

```
classDeclaration  ->
< modifier >  "class" identifier   [ "extends" className]   [ "implements"
interfaceName  <  "," interfaceName > ]   "{"  < fieldDeclaration >  "}"  .
modifier ->
"public"   | "private"   | "protected"   | "static"   | "final"   | "native"
| "synchronized"   | "abstract"   | "threadsafe"   | "transient"  .
identifier  ->
"a..z,$,_"  <  "a..z,$,_,0..9,unicode character over 00C0"  >  .
className  ->
identifier   | ( packageName  "." identifier )  .
interfaceName ->
identifier   | ( packageName  "." identifier )  .
```

```
fieldDeclaration  ->
(  [docComment]  ( methodDeclaration   | constructorDeclaration   |
variableDeclaration )  )   | staticInitializer   |  ";"  .
```


In Java, a class can define both the data structure and the algorithm for manipulating the data structure. The class name consists of an identifier (which can be of any length), and the class which may *extend* another class. In Java, classes can form an AKO (a-kind-of) taxonomy, as described in Chapter 1.

```
public class AppletFrame extends Frame {}
```

An instance of a class has a type. The *new* operator is used to make an instance of a class. For example:

```
class point {
     public double x,y;
}


point p1 = new point();
p1.x = 10;
p1.y = 11;
```

To gain access to the Java class libraries, you must import using an *import* statement. For example:

```
import java.awt.*;
import java.applet.Applet;
```

Import is described in more detail later in this chapter. To build on the methods and data structures of a parent class, a subclass is constructed that *extends* the parent class. For example:

```
public class AppletFrame extends Frame {
```

In this case, the *AppletFrame* class extends the *Frame* class. This means that the *AppletFrame* is a kind of *Frame*. Sometimes we would like to have a class that can never be instanced, only extended. We would declare the class as *abstract*. For example:

```
abstract class AppletUtil {
```

AppletUtil is an abstract class. We might like to keep *AppletUtil* abstract because it contains a combination of methods and fields that are to be inherited or overridden by a subclass. For example:

```
import java.applet.*;
import java.awt.*;


abstract class AppletUtil {
     static      Frame appletFrame = new Frame();


static void run ( Applet applet) {
           appletFrame.addNotify();
           appletFrame.add("Center",applet);
           appletFrame.resize(400,400);
           applet.init();
           appletFrame.show();
           applet.start();
           }
}
```

In this case, we can never make an instance of the *AppletUtil* class, because it is abstract. However, we can access the *appletFrame* variable and even invoke the *run* method without ever making an instance. For example, the following will run an instance of the *Applet* class, called *a*.

```
AppletUtil.run(a);
```

This example introduces the basic concept of a *method*. Classes in Java have two possible members: fields and methods. For example:

```
public class IHateHelloWorldExamples {
     public static void main(String args[]) {
          System.out.println("hello world");
     }
}
```

Here we see a class called *IHateHelloWorldExamples*. A Java application, *IHateHelloWorldExamples* contains a *main* method that is invoked at run time. This causes the "*hello world*" string to be printed on the console. The following, more elaborate example shows several numbered lines. The line numbers are for reference only. Lines 1 and 2 are used to import the Java class libraries.

```
1.      import java.util.*;
2.      import java.awt.*;
```

Line 3 shows one class subclassing (also known as extending) another class. Line 4 shows a class method with no return, not even *void*. This class method has the same name as the class and is called the *constructor*. The constructor returns an instance of the class when *new* is invoked.

```
3.      class Camera_grating_line extends Shape {
4.      Camera_grating_line() {
5.            color =
6.                Color.blue;
7.      }...
```

The *Camera_grating_line* class is a kind of *Shape*. It contains one method and no fields. The *color* field is stored in the *Shape* base class.

Overloaded Methods

As mentioned before, Java has no functions, only methods. Methods provide the algorithm for manipulating data that is stored in the class member variables. The constructor is a method. If no constructor is specified, a default constructor is provided. The default constructor takes *null* as an argument. The default constructor is overridden when another constructor is specified. For example:

```
class Lamp {
     boolean on;
     int Wattage;
     Lamp (int w) {
          Wattage = w;
     }
     Lamp ( ) {
     Wattage = 100;
```

```
    }
}
```

The *Lamp* constructor has been over loaded with two versions. The first version will support the following:

```
Lamp dim = new Lamp(40);
```

The second version supports the constructor invocation:

```
Lamp bright = new Lamp();
```

Java requires that each method have a unique signature. The signature of the method is determined by the number of arguments and their compile-time types.

Static Methods

Classes that contain static fields have only one copy (or instance) of the static member variable. The static member variable is therefore global to all instances of the class. For example:

```
import java.awt.*;
import java.applet.Applet;
class apple {
    static int numberOfApples = 0;
    apple() { numberOfApples++ ;}
}
```

Every time a new *apple* instance is made, the *numberOfApples* member will be incremented. Furthermore, all instances of the *apple* class will be able to access the same *numberOfApples* field. Thus, the *numberOfApples* int is stored in only a single place in memory.

Static members are instantiated when the class is initialized. They can not throw an exception without creating a compile-time error. Also, they can not refer to variables that have not been defined. The following code results in a compile–time error:

```
public class Lamp {
    static int wattage = voltage*current;
    static int voltage = 110;
```

```
    static int current = 1;
}
```


The error looks like this:

```
Error   : Can't make forward reference to voltage in class Lamp.
Lamp.java line 2     static int wattage = voltage*current;
```

On the other hand, if *power* is computed by a method, no such check is performed, although perhaps it should be. For example:

```
public class Lamp {
      static int wattage = power();
      static int voltage = 110;
      static int current = 1;
      static int power() {
           return voltage * current;
      }
       public static void main(String argv[]) {
            Lamp i = new Lamp();
            System.out.println("The power is "+ i.wattage);
       }

}
```

The preceding code prints the following:

```
The power is 0
```

Why should *power* be zero? *Power* is *current* times *voltage* and both are set to nonzero values. But, it compiles and runs. So what gives? The answer is the order of dependency. The *power* method is invoked first. The values for *voltage* and *current* are unset (and default to zero). Thus, *power* is set to the proper 110 watt answer only if both *current* and *voltage* are defined first. For example:

```
public class Lamp {
      static int voltage = 110;
      static int current = 1;
```

```
    static int wattage = power();
    static int power() {
         return voltage * current;
    }
     public static void main(String argv[]) {
          Lamp i = new Lamp();
          System.out.println("The power is "+ i.wattage);
     }

}
```

The preceding code prints the following:

```
The power is 110
```

There are two points to learn from this example:

1. Static variables must be defined in the order of independent first, dependent second.
2. Static methods cannot be used to avoid constraint 1.

Null

One of the literals of Java is *null*. It is what you get when nothing has been created. For example:

```
if (some_object != null) {
     System.out.println("Object Exists!");
}
```

Null has a null type and is the default value for any type that has not been created. For example:

```
class test {
     int i[99];
}
test foo = new test();
```

At this point, *foo.i* is equal to null. It is not until the memory is allocated for the array contained in the *foo* instance of the *test* class that the *i* field be non-null. For example the following code will set the *i* array, and *i* will no longer be null in value.

```
foo.i = new i[99];
```

Casting

Type conversion in Java is called *casting*. When casting is performed, it is a run-time operation. Casting converts only between compatible types.

Sometimes the only way to know for certain when types are compatible is to run the program. If a *ClassCastException* is thrown at run time, the type conversion failed.

It is always correct to cast an instance from a subclass to its superclass. For example:

```
1.     for (int i=0; i < v.size(); i++) {
2.                 s = (Shape) v.elementAt(i);
3.                 s.print();
4.                 }
```

In line 1, an instance of a vector, *v* is accessed for size. The elements in the vector are accessed using line 2. Note that each element in the vector is a class that extends the *Shape* class. It is always correct to cast the subclass of the *Shape* class back into the superclass. This technique enables the *print()* method to be invoked on each shape in the vector instance.

Subclassing and Super

One feature of the Java class is that it can intrinsically represent taxonomic structures (such as those described in Chapter 1). The taxonomic structures are formed by Java classes when a subclass *extends* a superclass. This type of extension is called *direct inheritance*. Thus, in terms of knowledge representation, Java classes can represent the AKO (a-kind-of) relationship. In addition, Java classes can represent the has-a relationship using the class member variables. For example, we can represent the statement "A student is a-kind-of human" by creating a *student* class that extends the *human* class. We can also represent the statement "The student has-a pencil" by placing a class member variable of *pencil* class type into the *student* class construct. In the following section we present the syntax of Java and its relationship to the semantics of Java.

A class can be used to provide a container for an instance variable of any primitive type. For example:

```
class Lamp {
     boolean on;
}
...
Lamp l = new Lamp ( );
l.on = true;
```

A Java class can be used to store a reference to named constants:

```
class Constants {
     static final double PIon2 = Math.PI / 2;
}
```

Notice that these class examples have no methods. When one class extends another, we are subclassing a superclass. The subclass will inherit the member variables, and methods, of the superclass. In the case of a name conflict, the subclass implementation always overrides the superclass implementation. For example:

```
class Lamp extends Constants {
     double power = 100 / PIon2; // watts
     boolean on = true;
     }
```

NOTE

The *power* in the *Lamp* class is set using a *PIon2* constant that is inherited from the *Constants* class. In this case, it is not strictly correct to say that the *Lamp* is a-kind-of *Constants*. Thus, the *extends* is being used as a programming convenience and not a means of knowledge representation.

On the other hand, consider this example:

```
class Student extends Human {
     Pencil p;
```

```
}
class Human {
     boolean bald = false;
}
```

Now we represent the statement “Doug is a bald student with a pencil”:

```
Student doug = new Student();
doug.p = new Pencil();
doug.bald = true;
```

Super is a keyword that permits a subclass to call upon the instance variable or method of the superclass. For example:

```
1.     public class ClosableFrame extends Frame {
2.     // constructor needed to pass window title to class Frame
3.     public ClosableFrame(String name) {
4.          // call java.awt.Frame(String) constructor
5.          super(name);
6.     }
7.     // needed to allow window close
8.     public boolean handleEvent(Event e) {
9.          // Window Destroy event
10.          if (e.id == Event.WINDOW_DESTROY) {
11.               dispose();
12.               return true;
13.          }
14.          // it's good form to let the superclass look at any unhandled
events
15.          return super.handleEvent(e);
16.     } // end handleEvent()
17.    } // end class ClosableFrame
```

Line 15 of class *ClosableFrame* invokes *super.handleEvent* because the event that was passed, *e*, may not have been an *Event.WINDOW_DESTROY* event. In that case, it may be that the superclass can decode and handle the event properly. A frame that extends the *ClosableFrame* will inherit the ability to handle the *Event.WINDOW_DESTROY* events by invoking *super.handleEvent(e).*

Abstract Classes and Methods

An abstract class is one that can never be instanced. It is a compile-time error to attempt to instantiate an abstract class. An abstract class must be extended before it can be instantiated. You can make a class abstract by declaring the class as abstract or by declaring a method within the class as abstract. Another way to declare an abstract class is to extend an interface or abstract class without providing an implementation for the abstract methods in the interface or superclass.

NOTE

A class should be declared abstract only if you intend to create subclasses to complete the implementation.

Abstract classes are used to create superclasses that can be extended. This situation typically involves the implementation of some, but not all of a class's methods. Methods without an implementation are declared abstract. Abstract methods must be implemented by a subclass before the subclass can be instanced. If a subclass does not implement the abstract methods, the subclass is also an abstract class even if it is not explicitly declared as abstract. Java supports polymorphism by casting instances into their common superclasses. The superclasses typically have abstract methods that are implemented by the subclasses. For example, in DiffCAD there is an abstract class called *Shape*:

```
abstract class Shape extends computation {
...
   abstract void draw(Graphics g);
...
```

The numerous kinds of *Shapes* in DiffCAD are stored in a vector instance called *drawnShapes.* Polymorphism is performed when the *Shape* subclass instances are accessed in the *drawnShapes* vector, cast into their *Shape* superclass, and then used as the target of the *draw* method. For example:

```
Shape s;
for (int i = 0; i < drawnShapes.size(); i++) {
```

```
        s = (Shape)drawnShapes.elementAt(i);
        s.draw(g);
    }
```


Shape polymorphism is possible for any of the methods (even the non-abstract ones) in the *Shape* class.

Final Classes and Methods

In contrast with abstract classes, in which a class must be subclassed to be extended, there are *final* classes. A class that is declared as final cannot be extended, although it can be instanced. Fields in a class that are declared as final become constant throughout the life of the class.

NOTE

To prevent a final class from being instanced, declare a single private constructor and never instance the class internally.

For example:

```
public final class Audio {

    // Prevent instantiation
    private Audio() {}
```

The *final* modifier prevents the *Audio* class from being extended. The *private* modifier makes the constructor invisible. Furthermore, the existence of a constructor will override the default constructor. Thus, no method outside the *Audio* class can instantiate the *Audio* class.

Final can also be used as a method modifier, preventing the method from being overridden. For example:

```
public final boolean keyDown(Event e, int key) {
...
```

This approach may lead to a compiler optimization as well as prevent any subclass from overriding the *keyDown* implementation.

When *final* is used on a class, type checking can occur at compile time.

Packages

The *package* statement is used to isolate the class and interface name space so that only public classes and interfaces will be used by nonpackage programmers. This technique restricts default access of classes and interfaces to package programmers.

The MBNF for packages follows:

```
packageStatement  ->
"package" packageName  ";"  .
packageName ->
identifier   |  ( packageName  "." identifier )  .
```

When you're building a large program, it is often to your advantage to divide it into subsections, each of which resides in a package. For example, DiffCAD has a Hypertext Markup Language (HTML) generator. HTML is a common file format that is read by browsers on the World Wide Web. The DiffCAD HTML generator reads C, C++ or Java code and outputs a colorized version of the source in HTML. The generator, which consists of 15 source files, has a single public class with a single public constructor. Anything that is not declared public in the package (i.e., anything that has private or default visibility) will not be visible even after the import statement is issued. The error

```
Error   : No constructor matching HtmlGenerator(java.lang.String) found in class
htmlconverter.HtmlGenerator.
process_menuitem.java line 21      new HtmlGenerator("The HTML Generator");
```

is emitted when

```
HtmlGenerator hg =
     new HtmlGenerator("The HTML Generator");
     hg.main(null);
```

is processed by the compiler, because the default visibility is not sufficient for access outside of the package. Only by providing the *public* accessor for the constructor can the HtmlGenerator class be instanced. For example:

```
public HtmlGenerator(String title) {
```

NOTE

A package permits a grouping of multiple classes and interfaces into a unit that can be developed and deployed separately.

The declaration of *public* from within a class is an explicit advertisement to those who would import your package. In effect, it says "You need this to use my package."

When classes are written into a source file with no package declaration, they become a part of the unnamed package. Code that resides in packages should not use code in an unnamed package. Separate compilation units are a design goal of Java developers, and packages are the mechanism for their creation.

Every class must reside in a compilation unit. An unnamed package is, in effect, an unnamed compilation unit. It is probably a Java design flaw to permit unnamed compilation units to compile.

Packaging is essential for writing large programs. Once upon a time, in a fit of quick prototyping frenzy, a programmer (who shall remain nameless) decided to write many classes without placing them in a package. The sophisticated reader will never do this. Everything worked fine, the days were sunny, and the programmer was productive. Then one day the program got so big and intertwined that even the programmer had trouble sorting out the interdependencies. "If only packaging had been enforced. Then I could have packaged as I went along," he thought. The sun began to set as his well-intentioned efforts to introduce packages into a large system began to fail. Even today, the programmer is trying to grapple with this massive code block. His productivity has fallen. The sun no longer shines. He has taken to drinking lots of java. His blood pressure has shot up as he became a Javaholic. This led to no good and he soon writing books on Java.

The moral of the story: Package small, package often!

Import

Import is a reserved keyword in Java that permits packages to be brought into the name space. Like libraries in C, packages are used to organize Java program libraries and to control type access. For example, a class or interface type is visible only outside a package if it is declared as *public*. The required Java packages are *java.lang*, *java.util* and *java.io*. The MBNF for the *import* statement follows:

```
 importStatement  ->
"import"  (  ( packageName  "."  "*"  ";"  )   |
( className  | interfaceName )  )  ";"  .
```

Extra *imports* are discarded but can make compilation take longer. For example:

```
import java.io.*;
import java.util.*;
```

Import all the public classes and interfaces in the *java.io* and *java.util* packages. It is possible to import a single class or interface from a package and thereby prevent a name–space conflict. For example:

```
import java.io.FilenameFilter;
import java.io.File;
```

These two classes are generally the only ones needed to build a *FilenameFilter*. You can build a package by the use of the *package* statement.

Visibility

Visibility is the aspect of an identifier that permits access to the name space that contains the identifier. If an identifier is visible, it may collide with another identifier. This is called name-space contention. Furthermore, if a method is not visible it cannot be invoked. The MBNF for the modifier is as follows:

```
modifier ->
"public"  | "private"  | "protected"  | "static"  | "final"  | "native"
"synchronized"  | "abstract"  | "threadsafe"  | "transient"  .
```

The visibility modifier is a subset consisting of the following

```
visibility_modifier ->
"public"  | "private"  | "protected" .
```

- *public*: gives the world access to your class, method, or field variable. It is required for those types you wish to advertise for use outside your package. The *public* modifier applies to classes, methods, and variables.
- *protected*: gives access to the class, subclass, and package. It can be applied only to methods and variables.

- *default*: gives package access. Classes that import the package will not be able to access types that have default access. To give such access, the type must be declared as public.
- *private protected*: gives class and subclass access. It can be applied only to methods and variables. *Private protected* prevents package access and is slated to be removed with JDK 1.1.
- *private*: gives access only within this class. Private access prevents inheritance. It can be applied only to methods and variables.

It is idiomatic in Java to have a collection of public static methods contained in a public final class with a private constructor that is never invoked. The *Math* package is like this, as is the DiffCAD file utility package called *futil*. For example:

```
public final class futil {
    /**
     * Don't let anyone instantiate this class.
     */
    private futil() {}
```

Table 2.2 shows the visibility matrix for Java.

TABLE 2. 2 VISIBILITY MATRIX

	Access by non-subclass from same package	**Access to non-subclass from different package**	**Access to subclass from different package**	**Inherited by subclass in same package**	**Inherited by subclass in different package**		**w=**
private	0	0	0	0	0	0	0
private protected	0	0	0	0	1	1	2
default	1	1	0	0	1	0	3
protected	1	1	0	0	1	1	4
public	1	1	1	1	1	1	6

A "1" in the visibility matrix indicates a true condition, a "0" indicates a false condition. The row is summed to obtain a “weight” for the visibility vector. The weight is shown in the *w* column. For example, *private* gives no access outside the class, so the weight is zero. *Public* gives you complete access, so the weight is 5. Here is an example of the use and abuse of the visibility modifiers:

```
public class visible {
      public String publicString = "publicString";
      String defaultString ="defaultString";
      protected String protectedString = "protectedString";
      private protected String privateProtectedString = "privateProtectedString";
      private String privateString = "privateString";
      // What follows is not in the Spec...it just worked here.
      private public String privatePublicString = "privatePublic";
     //gosh!
      public private  String publicPrivateString = "publicPrivate";
     //a stranger in a stranger land...
      protected public String protectedPublicString = "protectedPublic";
     //can you stand it?
      public protected String publicProtectedString = "publicProtected";
     //art is what you get away with.
      public protected private String publicProtectedPrivateString =
"publicProtectedPrivate";
      }

public class invisible extends visible {
      String iGetPublicStrings = publicString;
      String iGetDefaultStrings = defaultString;
      String iGetProtectedStrings = protectedString;
      String iGetPrivateProtectedStrings = privateProtectedString;
       // String iDontGetPrivateStrings = privateString;
       String iGetPrivatePublicStrings = privatePublicString;
       String iGetpublicPrivateStrings = publicPrivateString;
       String iGetProtectedPublicStrings = protectedPublicString;
       String iGetPublicProtectedStrings = publicProtectedString;
       String iGetPublicProtectedPrivateStrings = publicProtectedPrivateString;
      }
```

NOTE

The *public* modifier appears to override *private* no matter what the order.

This is not a part of the specification for Java. It just appears to work. It is probably within the vendor's right to declare a private public declaration a semantic error.

Interfaces

An interface is like an abstract class that has only abstract methods and constant fields. The interface can hold no method implementations and is defined like a class except that it uses the keyword *interface* rather than *class*. What follows is the MBNF for the *interface* declaration:

```
interfaceDeclaration  ->
  < modifier >  "interface" identifier   [ "extends" interfaceName  <  ","
  interfaceName > ]   "{"  < fieldDeclaration >  "}"  .
  fieldDeclaration  ->
  (  [docComment]  ( methodDeclaration   | constructorDeclaration   |
  variableDeclaration )  )   | staticInitializer   |  ";"  .
  interfaceName ->
  identifier   |  ( packageName  "." identifier )  .
```

Note that an *interface* declaration can extend multiple interfaces. The interface can serve as another reference type but can never be instanced. Thus, classes that implement an interface can always be cast back to the *interface* type.

Class identifiers and interface identifiers are always global to the package in which they reside. Therefore, classes and interfaces in the same package must have different identifiers. It is redundant to declare an interface as abstract, because interfaces are implicitly abstract. It is redundant to declare the field declarations as public, static, and final, because they are implicitly public, static, and final. All fields must be initialized. There is no primordial interface analogous to the *Object* class. It is a compile-time error to have ambiguous inherited fields in an interface. For example:

```
public interface real_dumb {
          double PI = 4;
     }
public interface dum_constants {
```

```
    double PI = 3;
}

public interface mixed_up_constants extends dum_constants, real_dumb
    {double foo=PI;}
Error   : Reference to PI is ambiguous. It is defined in interface real_dumb and
interface dum_constants.
constants.java line 17    {double foo=PI;}
```

The following are some correct uses of interfaces. The first example shows how interfaces can be used to group constants:

```
public interface constants {
     double Pi_on_180 = Math.PI / 180;
     double PI = Math.PI;
     double Pi_on_2 = Math.PI/2;
     double Pi_on_4 = Math.PI/4;
}
```

In the following example, a large array of symbols is being stored:

```
public interface CplusplusText {
    public static String cplusplusReservedWords[] = {
        "asm",
        "auto",
        "break",
        "case",
        "catch",
        "char",
        "class",
        "const",
        "continue",
        "default",
        "delete",
        "do",
        "double",
        "else",
        "enum",
        "extern",
```

```
        "float",
        "friend",
        "for",
        "goto",
        "if",
        "inline",
        "int",
        "long",
        "new",
        "operatcr",
        "private",
        "protected",
        "public",
        "register",
        "return",
        "short",
        "signed",
        "sizeof",
        "static",
        "struct",
        "switch",
        "this",
        "throw",
        "try",
        "typedef",
        "union",
        "unsigned",
        "virtual",
        "void",
        "volatile",
        "while"
    };
```

In the 125 files that make up our DiffCAD program, the interface was never used to extend multiple interfaces. We find that assertions in other books—that the interface gives Java a kind of multiple inheritance—are groundless. Java has no multiple inheritance for implementations. Java has multiple inheritance of prototypes via interfaces. A good use for multiple inheritance of prototypes is as a workaround for Java's strong typing, which constrains method invocation. We will see examples of this in Chapter 3.

Wrapper Classes

The wrapper classes are used to promote a primitive data type into a reference data type. After this promotion, the reference type permits a series of operations on the primitive data type that are not otherwise possible. Some things can be done only with an instance, such as the addition of an element to a vector. Primitive values cannot be the target of method invocations. All the wrapper types support a method invocation that converts them to a string representation, and all of them can be constructed from a string. All wrapper types support the following statement to check an object for equality of value:

```
public boolean equals(Object obj)
```

Keep in mind that the value of a wrapper object is stored in a private class field. When comparing two strings, use *equals* and never ==. For example:

```
if (arg.equals("this is right") {...
```

The preceding code will check the value of the string in *arg* against the value of "this is right". The following example checks the value of the *arg* reference with the compile-time constant string reference. This is probably not what the programmer wants:

```
if (arg == "this is wrong") {...
```

WARNING

Always use the *equals* method to check value when comparing wrapper instances.

Boolean

The *Boolean* class promotes the primitive *boolean* type to a reference type. Construction is overloaded to handle both the primitive *boolean* type and a string. The string must be equal to "true", ignoring case, before the *Boolean* instance will be true. For example:

```
Boolean b = Boolean("true");
Boolean b = Boolean("True");
Boolean b = Boolean("TRuE");
```


The preceding code lines will all lead to *b* representing true. Two static conversion methods are available for use in the *Boolean* class:

```
public static Boolean valueOf(String s) - returns a Boolean instance.
Boolean b = Boolean.valueOf("true"); // sets b to true
Boolean b = Boolean.valueOf("yes"); // sets b to false
```

NOTE This class is different from *getBoolean*, which has as its return a primitive *boolean* type. Note that sections 20.4.9 and 20.4.10 of the Java specification appear to show both of these functions returning the primitive *boolean*. This is an error in the specification.

```
public static boolean getBoolean(String name)
boolean b = Boolean.getBoolean("yes"); // sets b to false
```

Character

The *Character* class is a wrapper class for the primitive Java *char* data type. When embedded in the Java code, the 'a' character is surrounded by single quotes, whereas strings use double quotes. Any Unicode character is permitted as an argument for the construction of a *Character* instance. In addition, there are methods for converting to and from *Character* and *char*. For example:

```
Character aCharacter = new Character('π');
char aChar = '©';
a = new Character(aChar);
aChar  = aCharacter.charValue();
```

Neither an array of *char* nor an array of *Character* constitutes a *String*. The *Character* class supports several public static methods that are useful for testing and processing *chars*. All *chars* are Unicode, so all comparisons are written for Unicode. For example:

```
public static boolean isDigit(char ch)
```

```
return true if ch is a Unicode digit
```

Java version 1.0.1 supports ISO-LATIN-1 (0 through 9), Arabic-Indic, Extended Arabic-Indic, Devanagari, Bengali, Gurmukhi, Gujarati, Oriya, Tamil, Telugu, Kannada, Malayalam, Thai, and Lao digits.

The *Character* class has a series of public static *Char* test facilities that return a *boolean*. For example:

```
boolean aBoolean;
char aChar = 'a';

aBoolean =  Character.isDefined(aChar);
aBoolean =  Character.isLowerCase(aChar);
aBoolean =  Character.isUpperCase(aChar);
aBoolean =  Character.isTitleCase(aChar);
aBoolean =  Character.isDigit(aChar);
aBoolean =  Character.isLetter(aChar);
aBoolean =  Character.isLetterOrDigit(aChar);
aBoolean =  Character.isJavaLetter(aChar);
aBoolean =  Character.isJavaLetterOrDigit(aChar);)
aBoolean =  Character.isSpace(aChar);
```

The *Character* class also supports a series of *char* conversions that are public and static. These conversions can be accessed without making a *Character* instance. For example:

```
aChar =  Character.toLowerCase(aChar);
aChar =  Character.toUpperCase(aChar);
aChar =  Character.toTitleCase(aChar);

// returns an 'A', hex representation of 10.
aChar =  Character.forDigit(10, 16);

// returns 10, the hex value of 'A'
int digit = Character.digit(aChar, 16);
int i =     Character.MIN_RADIX // i = 2;
int i =     Character.MAX_RADIX // i = 36;
```

```
char aChar =     Character.MIN_VALUE // aChar = '\u0000';
char aChar = Character.MAX_VALUE // aChar = '\uffff';
```

About the only thing you can do with a *Character* instance is:

```
Character aCharacter = new Character('π');
String aString = aCharacter.toString();
aBoolean = aCharacter.equals(aCharacter);
int anInt = aCharacter.hashCode();
```

The Numeric Wrapper Classes

Each of the numeric primitive types has an associated wrapper class. In the following examples, let:

```
int i = 0;
long l = 0;
float f = 0;
double d = 0;
```

The name of the primitive numeric type, followed by its constructor wrapper class, follows:

```
Integer I =     new Integer(i);
Long L =     new Long(l);
Float F =     new Float(f);
Double D =     new Double(d);
```

Each of the numeric wrapper classes supports a conversion to a string:

```
String SI = I.toString();
String SL = L.toString();
String SF = F.toString();
String SD = D.toString();
```

Each of the numeric wrapper classes supports a conversion from a string to a reference type:

```
Integer I = new Integer(SI);
```

```
Long L = new Long(SL);
Float F = new Float(SF);
Double D = new Double(SD);
```

NOTE

Any string-to-numeric conversion can throw a *NumberFormat* exception. If you read a string (from a user's input or file), you are strongly advised to check the input.

Each of the numeric wrapper classes supports a conversion from a string to a primitive type using a static method invocation:

```
I = Integer.valueOf(SI);
L = Long.valueOf(SL);
F = Float.valueOf(SF);
D = Double.valueOf(SD);
```

An example of the use of the static *valueOf* method follows:

```
    try {value = Float.valueOf(getText()).floatValue();}
    catch(NumberFormatException e) {
      value = 0;
    }
```

The scalar numeric wrappers (*Integer* and *Long*) support an overloaded *valueOf* and *toString* method. The *toString* method will yield a string in any radix between 2 and 36, inclusive. For example:

```
SI = Integer.toString(i, 2); // yields a binary string
SL = Long.toString(l, 16); // yields a hex string
I = Integer.valueOf(SI,2); // converts from binary string
L = Long.valueOf(SL,16); // converts from hex string
```

To convert a string to a primitive, you can use the static methods *parseInt* and *parseLong*:

```
i = Integer.parseInt(SI); // converts from decimal string
i = Integer.parseInt(SI,2); // converts from binary string
```

```
l = Long.parseLong(SL,16); // converts from hex string
```


Strings

Java's *java.lang* package has a class known as *String*. Once they are created, *String* instances in Java cannot be changed. The reason is that once created, a *String* instance is added to an internal string pool. The pool is a private *HashTable* instance that speeds string lookup operations. We cannot find any public or private method for removal of a *String* instance from the internal hash table. Thus, once created, strings remain unchanged for the life of the Virtual Machine.

The *String* class supports a series of public static methods that can be accessed without making an instance of the *String* class. Suppose the following constants are predefined:

```
boolean aBoolean;
char achar;
int anInt;
long aLong;
float aFloat;
double aDouble;
Object anObject;
String aString;
char[] aCharArray;
int arrayOffset, numberOfElementsToConvert;
int endIndex; // last position in array - 1;
StringBuffer aStringBuffer = new StringBuffer(100);
Byte [] ASCIIByteArray;
int hiByte;
```

The following examples show how to use the *String* class, given the preceding, predefined contents:

```
aString = String.valueOf(anObject);
aString = String.valueOf(aCharArray);
aString = String.valueOf(aCharArray, arrayOffset, numberOfElementsToConvert);
aString = String.valueOf(aBoolean);
aString = String.valueOf(aChar);
```

```
aString = String.valueOf(anInt);
aString = String.valueOf(aLong);
aString = String.valueOf(aFloat);
aString = String.valueOf(aDouble);
```

The *String* class constructor is overloaded so that the following statements are true:

```
aString = new String();
aString = new String(anotherString);
aString = new String(aStringBuffer);
aString = new String(aCharArray);
aString = new String(aCharArray, arrayOffset, numberOfElementsToConvert);
```

The following examples show how to build *aString* with *hiByte* in the most significant eight bits, and *ASCIIByteArray* elements in the least significant eight bits (the *hiByte* is typically 0):

```
aString = new String(ASCIIByteArray, hiByte);
aString = new String(ASCIIByteArray, hiByte,
         arrayOffset,numberOfElementsToConvert);
```

The *String* class has a series of methods that use a string instance as a target. These methods typically involve extraction or conversion. For example:

```
aString = aString.toString();
aString = aString.substring(arrayOffset);
aString = aString.substring(arrayOffset, endIndex);
aString = aString.concat(anotherString)
//replace oldChar with aChar in aString
aString = aString.replace(oldChar, aChar);
aString = aString.toLowerCase();
aString = aString.toUpperCase();
// Trims leading and trailing whitespace
aString = aString.trim();
// intern returns a string from the private internal

// hash table set that will pass the

// aString == anotherString test
```

```
aString = anotherString.intern();
```



```
aBoolean = aString.equals(anObject);
anInt = aString.hashCode();
andInt = aString.length();
aChar = astring.charAt(anInt);
// get charArray from aString
aString.getChars(beginIndex,endIndex, charArray, arrayOffset);
aString.getBytes(beginIndex,endIndex, byteArray, arrayOffset)
charArray = aString.toCharArray();
aBoolean = aString.equalsIgnoreCase(anotherString);
```

CompareTo is a method that returns –1, +1, or 0. The string comparison is performed using the underlying Unicode characters.

```
anInt = aString.compareTo(anotherString)
aBoolean = aString.regionMatches(beginIndex,anotherString,
                arrayOffset, numberOfElementsToConvert)
aBoolean = aString.regionMatches(ignoreCase, arrayOffset,
     anotherString, arrayOffset, numberOfElementsToConvert)
aBoolean = aString.startsWith(aString)
aBoolean = aString.startsWith(aString, arrayOffset)
aBoolean = aString.endsWith(aString)
```

There are some string search functions in the *String* class. When they fail, they return –1.

```
// first location of aChar in aString

anInt = aString.indexOf(aChar);
// first location of aChar in aString starting from anInt
anInt = aString.indexOf(aChar, anInt);
// first location of aString in anotherString
anInt = anotherString.indexOf(aString);
// first location of aString in
// anotherString starting from anInt

anInt = anotherString.indexOf(aString, anInt);
```

```
// Find the last occurrence of a string or char in a string

// starting from anInt
anInt = aString.lastIndexOf(aChar);
anInt = aString.lastIndexOf(aChar, anInt);
anInt = aString.lastIndexOf(aString)
anInt = aString.lastIndexOf(aString, anInt);
```

Arrays

An instance of an array is a reference-type instance. The *length* variable can be accessed, but not set, and it indicates the number of elements in the array. Array elements start at element zero and end at element length –1.

Any access beyond the end of the length of an array results in the *ArrayIndexOutOfBoundsException* being thrown. The method *arraycopy* will copy values between arrays of the same type.

An array can hold any data type, but every element in the array must hold the same data type. Arrays have a fixed length and can be dynamically allocated. As long as there is enough memory, it is possible to make deep arrays. For example:

```
        int deep_array[] [] [] [] [] [] [] [] [] [] = new int [2] [2] [2] [2]
[2] [2] [2] [2] [2] [2]  ;
        int sum = 0;
        for (int i=0; i<2; i++) {
            deep_array[i] [i] [i] [i] [i] [i] [i] [i] [i] [i] = 1;
            sum +=  deep_array[i] [i] [i] [i] [i] [i] [i] [i] [i] [i];
            System.out println("sum =" + sum);
         }
sum =1
sum =2
```

As pointed out in Chapter 1, there are both Java-style arrays and C/C++-style arrays.

Vectors

A Vector is a container class that can be expanded dynamically. A container class is designed to store reference data types. Vectors are an excellent choice for implementing data structures of variable length. The *Vector* class resides in the *java.util* package. The *java.util* package must be imported before using. Vectors can hold any combination of objects. For example:

```
Vector v = new Vector();
Rectangle a = new Rectangle(10,20);
v.addElement(A);
```

Vectors cannot hold *int*, *float*, *char*, *byte*, or any of the nonreference data types. The following example shows a series of different data types being added to a *Vector* instance:

```
static public void add_elements(Vector v) {
 for (int i=0; i < number_of_rays.getValue(); i++) {
      v.addElement(grating_target_line[i]);
        v.addElement(camera_grating_line[i]);
   }
     v.addElement(camera);
     v.addElement(grating);
     v.addElement(wedge);
     v.addElement(laser);
     v.addElement(laser_wedge_line);
} // add_elements
```

Having a collection of different instances in a *Vector* instance is a good way to implement polymorphism. With polymorphism, you can cast all the instances stored in the *Vector* instance into a common superclass that supports a common method. In the following example, *drawShapes* is a vector that contains several different instances, all of them subclasses of the *Shape* class. Thus, all of them can be cast into the *Shape* type. Also, the *Shape* class has an abstract *draw* method. Thus, the *draw* method can be invoked on any subclass of the *Shape* class provided that it has been properly cast. For example:

```
for (int i = 0; i < drawnShapes.size(); i++) {
```

```
        s = (Shape)drawnShapes.elementAt(i);
        s.draw(g);
    }
```

Exceptions

The grammar of a computer language allows the formulation of code that will compile. However, after compile time comes run time. The Java language specification covers both the syntactic (compile-time) and the semantic (run-time) behavior of the Java environment. Thus, the specification goes beyond the compiler and describes the behavior of the Java virtual machine. The Java virtual machine will throw an exception under special run-time conditions. This throw is a nonlocal transfer of the flow of control from the area that generated the exception to a place that *catches* the exception.

The MBNF of the *try* statement follows:

```
tryStatement ->
"try" statement   <  "catch"  "(" parameter  ")" statement >
[ "finally" statement]  .
```

Typically, a statement that can possibly fail is surrounded with the *try* and *catch* keywords. Failure in the statement following the *try* can then be intercepted and handled. For example:

```
try {
              // create an instance of your applet class
              a = (Applet) Class.forName(className).newInstance();
         } catch (ClassNotFoundException e) {
              System.out.println("ClassNotFoundException in AppletFrame");
              return;
         } catch (InstantiationException e) {
              System.out.println("InstantiationException in AppletFrame");
              return;
         } catch (IllegalAccessException e) {
              System.out.println("IllegalAccessException in AppletFrame");
              return;
         }
```

Sometimes no error is to be generated at all; we simply want the program to continue running. For example:

```
try {
  out.write(buffer);
  } catch(Exception e) { }
```

You can invent exception names at will by extending the *exception* class. For example:

```
class FileFormatException extends Exception {
    public FileFormatException(String s) {
        super(s);
    }
```

An overview of the *java.lang* exception classes is shown in Figure 2.3.

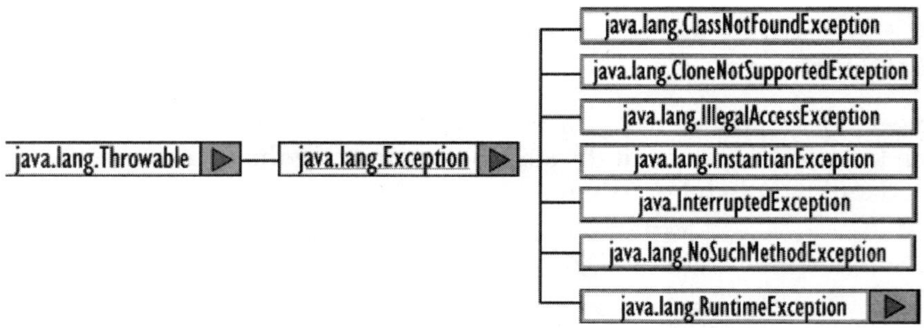

FIGURE 2.3. AN OVERVIEW OF THE JAVA.LANG EXCEPTION CLASSES.

The *java.lang.RuntimeException* class hierarchy is shown in Figure 2.4.

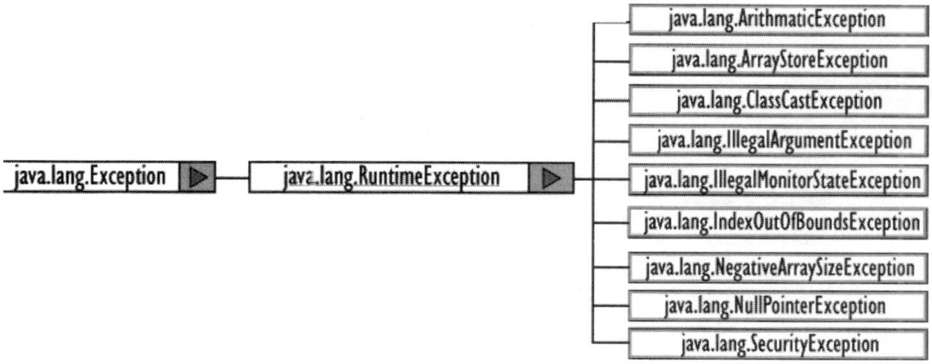

FIGURE 2.4. AN OVERVIEW OF THE JAVA.LANG.RUNTIME EXCEPTION CLASSES.

Threads

Threads enable parallelism within a program. Threading provides a low-overhead context switch. In the DiffCAD program, we use threads to display a digital clock in the lower-left corner of the screen, shown in Figure 2.5.

FIGURE 2.5. THE DIGITAL CLOCK.

The Java specification says that threads use a shared memory paradigm to perform interthread communication. This usage is typical, but threads can use any number of techniques to perform communication. For example, you could simulate a distributed computation scheme using threads and another communication technique, perhaps involving the socket API. The Java language specification also says that "threads may be supported by having many hardware processors." In fact, this capability is not yet implemented on any virtual machine that we know of, although we expect this situation to change. Java uses the *synchronized* statement to make sure that a method contained within a thread is executed atomically. *Atomic* operations appear to happen all at once. Thus, synchronized methods cannot be interrupted by other threads. To put it another way, *synchronized* methods prevent other threads from running. If the synchronized method takes longer than the time provided by a sleep method invocation, the *synchronized* method may not be updated frequently enough. The *Thread* class hierarchy is shown in Figure 2.6.

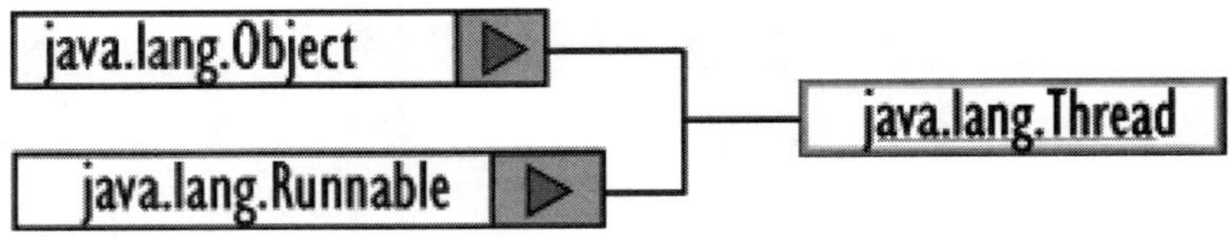

FIGURE 2.6 THE THREAD CLASS HIERARCHY

Synchronized methods allow only one thread to access data at a time, this can prevent data corruption. Synchronized methods can also prevent *race* conditions, which arise when different threads have access to the same variable in an instance. If both threads access the instance variable at the same time, it is a "race" to see which thread will effect the change first, (hence the term). The solution is to create synchronized accessor methods for every variable in an instance. Then any change made through the synchronized accessor methods will be atomic. If there is a mix of synchronized accessor methods and asynchronous accessor methods, the alterations to the instance variables' values are not atomic and the race condition may still exist. Threads are run with a round-robin, priority-queue–driven scheduler. This arrangement can be overridden with custom schedulers (whose composition is beyond the scope of this text). Furthermore, if the Java virtual machine is written that takes advantage of multiple CPUs, task completion on multiple threads will not be predictable.

To make a new thread, subclass the *Thread* class and implement the *run()* method. Instantiate the thread subclass and then invoke the *start()* method on the instance to begin thread execution.

Another way to make a thread is to use the *Runnable* interface. Any class that implements the *Runnable* interface must implement the *run* method, which is invoked when the thread is started. The thread is started when an instance of the *Runnable* class is passed to the *Thread* constructor and *start* is invoked.

The following is a simple thread example called *RaceThread.*

NOTE

The *RaceThread* class extends the *Thread* class. To run an instance of *RaceThread*, you must make an instance of *RaceThread* and use this instance as the target of a *run()* invocation.

```
1.     class RaceThread extends Thread {

2.         public void run() {
3.                 while (true) {
4.                     System.out.println("Priority=\t" + getPriority());
5.                     System.out.println("toString=\t"+toString());
6.                     System.out.println("getName=\t"+ getName());
7.                     System.out.println("isDaemon=\t"+isDaemon());
8.                     System.out.println("isAlive=\t"+isAlive());
9.                     try {Thread.sleep(10000);}
10.                     catch (InterruptedException e) {}
11.                     }
12.          }
13.    }
```

Take a close look at the *run* method. The *run* method is at the heart of any thread's activity. Classes that extend the *Thread* class must implement the *run* method or they will not run (they compile, but upon launch nothing happens). Next, look at the *while* statement on line 3. It is an infinite loop! It will run until terminated. This makes line 9 and 10 crucial. Line 9 will put the thread to sleep for 10,000 milliseconds (10 seconds). The argument to *Thread.sleep* must be a long integer. This makes sense when you consider that there are

$$2^{\log_2(1000*60*60*24)} = 2^{26}$$

milliseconds in a day. To put it another way, a signed 32–bit integer will overflow in

$$2^{31} - 1$$

milliseconds, which is slightly more than 3.5 weeks. A 64–bit integer will overflow in 292.4 billion years. At that point we will probably not be using 64–bit integers (or even FORTRAN). To put this in perspective, the sun's corona will have engulfed Earth by then, and this book will almost certainly be out of print. To run the *RaceThread* example, use the following code:

```
RaceThread r = new RaceThread();
r.start();
```

The following will be printed every 10 seconds at the console:

```
Priority=     5
toString=     Thread[Thread-2,5,main]
getName=     Thread-2
isDaemon=     false
isAlive=     true
```

Java makes threading a part of the *java.lang* package. This means that you can expect threading to be implemented even on an embedded picoJava controller. Having threading described in the Java specification means that multithreaded Java programs are as portable as Java.

The following code creates the digital clock in DiffCAD:

```
1.     import java.util.*;
2.     import java.awt.*;
3.     import java.applet.Applet;
```

Lines 1, 2, and 3 import the packages that DiffCAD needs in the *DigitalThread* class.

```
4.     // DigitalThread must extend applet to getFontMetrics.
5.     public class DigitalThreads  extends Applet implements Runnable {
```

A class that implements *Runnable* must provide an implementation for the *run* method. Note that the *Thread* class has a specific type:

```
6.     Thread runner;
```

The Java machine will execute threads until they die. A thread is dead when its *run()* method returns or the *stop()* method is invoked.

```
7.      Graphics g;
8.      Frame f;
9.      public void start() {
10.           if (runner == null) {
11.                runner = new Thread(this);
12.                runner.start();
13.           }
14.     }
```

The *stop* method allows you to invoke the *stop* method of a thread that has already been stopped. You may need to stop a thread in order to avoid a race condition or to make sure that system resources are made available for other threads. After the *stop()* method is invoked, it may be that the thread should be restarted. Lines 9–14 enable you to start a thread that has been stopped. You can invoke *start* on a method that is already started.

```
15.     public void stop() {
16.          if (runner != null) {
17.               runner.stop();
18.               runner = null;
19.               }
20.     }
```

In line 21, note the use of the *synchronized* keyword. This makes the *draw* method atomic. *Draw* is one method we do not want to interrupt, particularly with another *draw* method. Should this occur, *draw* could leave the screen in an inconsistent state. The frame field, *f*, is set by the calling program. On line 25 we see that the date is being accessed from the *java.util* API.

```
21.     synchronized private void draw() {
22.               Dimension dim = f.size();
23.               int height = dim.height - 60;
24.               int width = dim.width;
25.                  Date theDate = new Date();
26.                  String date_string = theDate.toString();
```

A notable problem with the *theDate.toString()* method is that it will yield different returns on different platforms. For example, on Microsoft's' J++ compiler, running under Windows 95/NT we found that the Eastern Daylight Time or Standard Time (EDT, ST) initials are present. For this graphics routine to work, we must therefore query the graphics context for the font metrics, to size a background–fill rectangle properly.


```
27.             int xloc = 10;
28.             int yloc = dim.height - 60;
```

The upper–left corner of the screen is the origin in Java. Thus, (dim.width, dim.height) are the coordinates of the lower–right corner.

```
29.           // g.setFont(f);
30.             // get the string width in pixels.
31.             int string_width = getFontMetrics(
32.                 g.getFont()).stringWidth(date_string);
33.             int string_height = getFontMetrics(
34.                 g.getFont()).getHeight();
```

Line 32 uses the graphics context, *g*, which is set using the calling program. Thus, both the graphics context and the frame for the clock must be given before this program will have a draw target. Line 35 draws a filled clear rectangle under the string so that the string can be seen clearly even if there is clutter on the screen.

```
35.             g.clearRect(xloc,yloc,string_width,string_height);
36.           g.drawString(date_string, xloc,height+xloc);
37.     }
```

On line 38 we see the *run* method. The *run* method is required of classes that implement the *runnable* interface. Note that the only synchronized method, *draw*, is invoked on line 40. This method gives us a return within a second, which is the sleep time for this thread, as seen on line 41.

```
38.     public void run() {
39.         while (true) {
40.             draw();
41.             try {Thread.sleep(1000);}
42.             catch (InterruptedException e) {}
43.         }
44.     }
45.     }
```

Start the *DigitalThreads* class by making an instance, setting the frame and graphics variables, and then invoking the *run* method:

```
1.        public static DigitalThreads
2.              clock_thread = new DigitalThreads();
3.    void start_clock() {
4.        clock_thread.g = getGraphics();
5.        clock_thread.f = main_frame;
6.        clock_thread.start();


7.    }
```

Note that in line 4 the graphics context is obtained from the local graphics context. This assumes that the graphics context for the instance of the class exists. Also, note that a *main_frame* variable was available for use in line 5. In fact, the clock could be placed in any frame and graphics context. Notice that the clock will update the main frame even if it becomes hidden by other frames. The clock thread does not check when its frame is hidden.

Thread Groups

One of the issues facing programmers who want to create many threads is that of limited system resources. CPU time is a limited resource, and we need to keep an eye on the CPU usage and space requirements of the started threads. This information is critical, and it is needed by both the program designer and the program itself. The designer requires this information in order to engineer a program that reacts well to the changing load on a system.

Every thread belongs to a *thread group*. The thread groups are arranged in a thread group hierarchy, at the root of which is the *system thread group*. The system thread group consists of four threads: the garbage collector, the clock handler, the idle thread, and the finalizer thread. The *main group*, a child of the system thread group, has the default thread that starts at the *main()* method. When windows are started, new threads are added to the main group. These threads include, but are not limited to, the AWT input thread, AWT toolkit thread, and screen updater thread.

The Thread Manager

This section shows how to write a program that will enable the display of the various threads running on a system. At the heart of the thread management concept is the idea that we need to access the system thread group (i.e., the root thread) and then list all the children. The procedure is to create a thread and then traverse up the parent groups until we reach the root group, which is identified by its null parentage. The following code returns a reference to the current thread.

```
Thread.currentThread()
```

This code returns a reference to the group that the current thread belongs to:

```
Thread.currentThread().getThreadGroup()
```

The reference is of *ThreadGroup* type. Once we have an instance of the current thread group, we can obtain the parent of the current thread group, which is also a thread group. We continue this traversal until the parent of the current thread group is null.

The following routine returns the *SystemThreadGroup*:

```
public ThreadGroup getSystemThreadGroup() {


    ThreadGroup systemThreadGroup;
    ThreadGroup parentThreadGroup;


    systemThreadGroup = Thread.currentThread().getThreadGroup();


    while ( ( parentThreadGroup = systemThreadGroup.getParent()) != null)
        systemThreadGroup = parentThreadGroup;


    return systemThreadGroup;
}
```

The following code returns the groups of the system in an array of groups. We note that the number of groups is increased by 1 so that the root group can be added to the end of the array:

```
public ThreadGroup[] getThreadGroupsArray() {
```

```
2.          ThreadGroup systemThreadGroup = getSystemThreadGroup();
3.          int numberOfGroups = systemThreadGroup.activeGroupCount() +1;
4.          ThreadGroup threadGroupsArray[] = new
5.               ThreadGroup[numberOfGroups];
6.          systemThreadGroup.enumerate(threadGroupsArray);
7.          threadGroupsArray[numberOfGroups] = systemThreadGroup;
8.          return threadGroupsArray;
9.     }
```

When *group_instance.enumerate(ThreadGroup list[])* is called, all the thread group instances and their descendants are placed into the list. The list does not include the *group_instance* itself. This is why it is added to the end of the array on line 7.

Another form of *enumerate* is associated with a thread instance. *group_instance.enumerate(Thread list[])* returns an array of threads in this thread group and all descendant thread groups.

```
java.lang.ThreadGroup[name=system,maxpri=10]
    Thread[Finalizer thread,1,system]
    Thread[Idle thread,1,system]
    java.lang.ThreadGroup[name=main,maxpri=10]
        Thread[main,5,main]
        Thread[Thread-2,5,main]
        Thread[AWT-macos,5,main]
        Thread[Thread-3,5,main]
        Thread[Image Fetcher 0,8,main]
        Thread[Image Fetcher 1,8,main]
        Thread[Image Fetcher 2,8,main]
        Thread[Image Fetcher 3,8,main]
        Thread[Screen Updater,4,main]
        Thread[Audio Player,10,main]
```

The entire thread, called *RaceThread* (the new and improved version), follows:

```
class RaceThread extends Thread {
         public void run() {
```

```
                while (true) {
                    printThreadGroups();
                    try {Thread.sleep(10000);}
                    catch (InterruptedException e) {}

                    }
            }

public ThreadGroup getSystemThreadGroup() {
     ThreadGroup systemThreadGroup;
     ThreadGroup parentThreadGroup;


     systemThreadGroup = Thread.currentThread().getThreadGroup();
     while ( ( parentThreadGroup = systemThreadGroup.getParent()) != null)
          systemThreadGroup = parentThreadGroup;
     return systemThreadGroup;
}

public ThreadGroup[] getThreadGroupsArray() {
          ThreadGroup systemThreadGroup = getSystemThreadGroup();
          int numberOfGroups = systemThreadGroup.activeGroupCount() +1;
          ThreadGroup threadGroupsArray[] = new
               ThreadGroup[numberOfGroups];
          systemThreadGroup.enumerate(threadGroupsArray);
          threadGroupsArray[numberOfGroups] = systemThreadGroup;
          return threadGroupsArray;
     }


public Thread []  getThreadsArray() {
          ThreadGroup systemThreadGroup = getSystemThreadGroup();
          Thread threadsArray[]= new Thread[systemThreadGroup.activeCount()];
           systemThreadGroup.enumerate(threadsArray);
          return threadsArray ;
     }


     public void printThreadGroups() {
          getSystemThreadGroup().list();
```

```
    }

    public void printThreads() {
        Thread [] threadsArray = getThreadsArray();
        for (int i = 0; i < threadsArray.length; i++)
            System.out.println(threadsArray[i]);
    }

    public void printThreadsAndGroups() {
        System.out.println("The threads are:");
        printThreads();
        System.out.println("The groups are:");
        printThreadGroups();
    }

}
```

Summary

This chapter introduced some Java programming basics, although a great deal of Java was not covered. The Java API, which generally fills a large reference book, was scarcely touched. Several excellent reference books on Java exist, and we wholeheartedly suggest that you read a few of them.

Roasting Your Own Java

Yield: 1/4 lb of the finest French roast I ever had.

Place 113 g (about 1/4 lb) of Kenya AA green coffee

into a Presto PopCornNow™ Plus air popper.

For French Roast, roast for 5 minutes.

WARNING

Never leave roaster unattended. Always roast outdoors, because roasting makes a heavy blue smoke. Replace the plastic hood with a 7.6 cm (3–inch) diameter by 20.3 cm (8–inch) metal stove pipe to prevent melting and discoloration of the hood. Stove pipes are available in home centers. Roasting with a hot-air popper violates the warranty and makes popcorn taste funny.

Chapter 3

The Graphical User Interface

This chapter provides an overview of Java's graphic user interface. Central to the look and feel of the Java GUI is the AWT (Abstract Window Toolkit). The AWT consists of a collection of classes, some of which are extended six or more levels deep. For example, a programmer who extends an *Applet* class inherits the methods and instance variables of *Object*, *Component*, *Container*, and *Panel* superclasses. For us to cover the *Applet* class would mean also describing the four base classes, something that is better left to references such as [Chan and Lee]. Our approach is to cover the material that is essential to understanding how to write digital signal processing programs. The AWT is rich with many GUI options and widgets, but we will cover only those that are essential to the task. The finer points of the AWT are better left to book dedicated to that subject, such as [Geary and McClellan].

We begin with the *Color* class and then turn to the *Graphics* class. For the purpose of digital signal processing we need to learn how to draw a line, a disk, a string, and an image and how to set colors and fonts. The *Graphics* class is deep, and we do not attempt to cover it completely.

The Color Class

The *Color* class is a public final class that is based on the RGB color model. The class has several public final static colors that are predefined. You can specify an instance of the *Color* class using two variants of the additive color synthesis system: RGB (red, green, and blue) or HSB (hue, saturation, and brightness). The additive color synthesis system is based on the premise that black is the absence of color and white is the combination of all colors. For example, when red, green, and blue are combined in equal amounts, the resulting color is white. This system is consistent with electronic displays that emit light. The subtractive synthesis color system is based on the premise that black is the combination of all colors and white is the absence of color. This is a result of light absorbing pigments that are applied to a white background. The subtractive synthesis color primaries are CMY (cyan, magenta, and yellow). The AWT does not support subtractive synthesis.

Class Summary

```
public final class Color {
public final static Color white
public final static Color lightGray
public final static Color gray
public final static Color darkGray
public final static Color black
public final static Color red
public final static Color pink
public final static Color orange
public final static Color yellow
public final static Color green
public final static Color magenta
public final static Color cyan
public final static Color blue
public Color(int r, int g, int b)
public Color(int rgb)
public Color(float r, float g, float b)
public int getRed()
public int getGreen()
public int getBlue()
public int getRGB()
public Color brighter()
public Color darker()
public int hashCode()
public boolean equals(Object obj)
public String toString()
public static Color getColor(String nm)
public static Color getColor(String nm, Color v)
public static Color getColor(String nm, int v)
public static int HSBtoRGB(float hue, float saturation, float brightness)
public static float[] RGBtoHSB(int r, int g, int b, float[] hsbvals)
public static Color getHSBColor(float h, float s, float b)
}
```

Class Usage

Suppose that the following variables are defined:

```
Color c,c1;
int r, g, b;
float rf, gf, bf;
int anInt;
```

To access the built-in color names, we use the following:

```
c = Color.white;
c = Color.gray;
c = Color.lightGray;
c = Color.darkGray;
c = Color.black;
c = Color.red;
c = Color.pink;
c = Color.orange;
c = Color.yellow;
c = Color.green;
c = Color.magneta;
c = Color.cyan;
c = Color.blue;
```

To construct a new color, we use this:

```
c = new Color(r, g, b);
```

The AWT assumes that the *r*, *g*, and *b* quantities range from 0 to 255. If the range assumption is violated, the value is truncated. In fact, the color is typically a packed *int* consisting of a blue component in bits 0–7, a green component in bits 8–15, and a red component in bits 16–23. To instantiate a new color from an existing instance, we use this statement:

```
c1 = new Color(c);
```

To instantiate a new color from red, green and blue components in the range from 0 to 1, inclusive, we use this:

```
c = new Color(rf, gf, bf);
```

To get the components of a color, we use this:

```
r = c.getRed();
r = c.getGreen();
r = c.getBlue();
```

To get a 24–bit RGB color packed into an *int* and typed as an *int*, we use this:

```
anInt = c.getRGB();
```

To brighten or darken a color:

```
c.brighter();
c.darker();
```

To obtain the color *hashCode* (implemented as *getRGB()*):

```
anInt = c.hashCode();
```

To compare color values:

```
aBoolean = c.equals(c2);
```

To convert to a string:

```
str = c.toString();
```

If *nm* is a valid color property name, the *Integer* class can use the *getInteger* method to map the property name into a color. The following static methods are used:

```
c1 = Color.getColor(nm);
```

To return a color if the property name is undefined:

```
c1 = Color.getColor(nm, aColor);
```

To return a color based on a 24–bit RGB color *int* if the name is undefined:

```
c1 = Color.getColor(nm, anInt);
```

NOTE

Color properties must be set by the Java application. They are not predefined.

To convert a floating–point description of hue, saturation, and brightness to the RGB model (with a 24–bit *int*) use the following:

```
// ranging from 0..1 inclusive
float hue, saturation, brightness;
anInt = Color.HSBtoRGB(hue, saturation, brightness);
```

To convert from RBG to HSB, there are two methods to choose from:

```
float hsbvals[];
int r, g, b; // r, g and b range from 0..255
hsbvals = Color.RGBtoHSB(r, g, b, null);
Color.RGBtoHSB(r, g, b, hsbvals);
```

Both techniques return three values in the *hsbvals* array.

To make an instance of a color from an HSB floating–point triplet, use the following:

```
Color aColor = Color.HSBtoRGB(h, s, b);
```

The Graphics Class

The *Graphics* class is used to draw lines, shapes, images, and characters. The *Graphics* class resides in the *java.awt* package and is an abstract base class. Because it is abstract, it cannot be created by the invocation of *new*. Instead, it is created by an instance of a *Component*. The *Graphics* class has a default coordinate system, as shown in Figure 3.1.

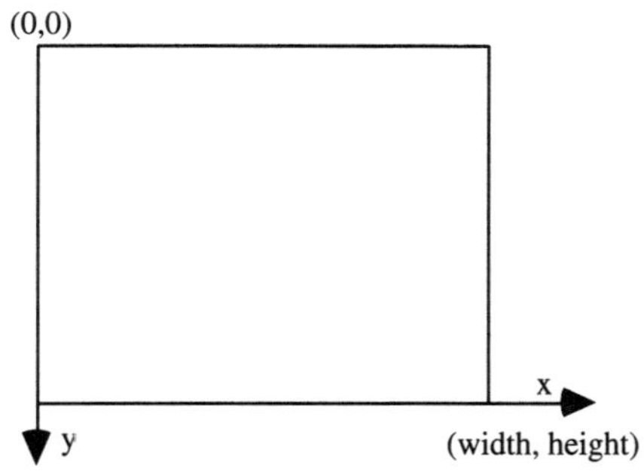

FIGURE 3.1 THE DEFAULT COORDINATE SYSTEM OF THE GRAPHICS CLASS.

Class Summary

```
public abstract class Graphics {
public abstract Graphics create()
public Graphics create(int x, int y, int width, int height)
public abstract void translate(int x, int y)
public abstract Color getColor()
public abstract void setColor(Color c)
public abstract void setPaintMode()
public abstract void setXORMode(Color c1)
public abstract Font getFont()
public abstract void setFont(Font font)
```

```
public FontMetrics getFontMetrics()
public abstract FontMetrics getFontMetrics(Font f)
public abstract Rectangle getClipRect()
public abstract void clipRect(int x, int y, int width, int height)
public abstract void copyArea(int x, int y, int width, int height, int dx, int dy)
public abstract void drawLine(int x1, int y1, int x2, int y2)
public abstract void fillRect(int x, int y, int width, int height)
public void drawRect(int x, int y, int width, int height)
public abstract void clearRect(int x, int y, int width, int height)
public abstract void drawRoundRect(int x, int y, int width, int height, int
arcWidth, int arcHeight)
public abstract void fillRoundRect(int x, int y, int width, int height, int
arcWidth, int arcHeight)
public void draw3DRect(int x, int y, int width, int height, boolean raised)
public void fill3DRect(int x, int y, int width, int height, boolean raised)
public abstract void drawOval(int x, int y, int width, int height)
public abstract void fillOval(int x, int y, int width, int height)
public abstract void drawArc(int x, int y, int width, int height,int startAngle,
int arcAngle)
public abstract void fillArc(int x, int y, int width, int height,int startAngle,
int arcAngle)
public abstract void drawPolygon(int xPoints[], int yPoints[], int nPoints)
public void drawPolygon(Polygon p)
public abstract void fillPolygon(int xPoints[], int yPoints[], int nPoints)
public void fillPolygon(Polygon p)
public abstract void drawString(String str, int x, int y)
public void drawChars(char data[], int offset, int length, int x, int y)
public void drawBytes(byte data[], int offset, int length, int x, int y)
public abstract boolean drawImage(Image img, int x, int y,  ImageObserver
observer)
public abstract boolean drawImage(Image img, int x, int y, int width, int height,
ImageObserver observer)
public abstract boolean drawImage(Image img, int x, int y,  Color bgcolor,
ImageObserver observer)
public abstract boolean drawImage(Image img, int x, int y, int width, int height,
Color bgcolor, ImageObserver observer)
public abstract void dispose()
public void finalize()
public String toString()
}
```

Class Usage

The *Graphics* class supports a simple raster graphics package. Suppose the following constants are predefined:

```
Color aColor;
Graphics g;
Rectangle aRectangle;
int x, y, x1, y1, x2, y2;
Boolean raised;
int height, width, arcHeight, arcWidth, nPoint;
int xArray[], yArray[];
char charArray[];
int numberOfItemToDraw, offsetIntoArray;
byte byteArray[];
Image img;
ImageObserver anImageObserver;
```

To create a graphics instance, use the following:

```
Graphics g1 = g.create(x, y, width, height);
```

There are color methods for getting and setting the foreground color of the *Graphics* instance:

```
aColor = g.getColor();
g.setColor(aColor);
```

There are paint methods for setting the *Graphics* instance to overwrite in xor paint mode:

```
g.setPaintMode();
g.setXORMode(aColor);
```

There are clipping methods for shrinking and for getting the clipping rectangle:

```
aRectangle = g.getClipRect();
g.clipRect(x, y, width, height);
```

The *Graphics* class supports a method for painting a rectangle with the background color:

```
clearRect(int x, int y, int width, int height);
```

To translate the coordinates of the *Graphics* instance, use this:

```
g.translate(x, y);
```

To get and set the fonts of the *Graphics* instance:

```
aFont = g.getFont();
g.setFont(aFont);
```

To get the font metrics for a *Graphics* instance or for a font:

```
theFontMetrics = g.getFontMetrics();
theFontMetrics = g.getFontMetrics(aFont);
```

To get the clipping rectangle:

```
aRectangle = g.getClipRect();
```

To shrink the clipping rectangle:

```
g.clipRect(x, y, width, height);
```

To copy and translate a rectangular area of the screen by (dx, dy):

```
g.copyArea(x, y, width, height, dx, dy);
```

To draw a line between (x1, y1) and (x2, y2):

```
g.drawLine(x1, y1, x2, y2);
```

To fill a rectangle with the current color:

```
g.fillRect(x, y, width, height);
```

To draw a rectangle outline with the current color:

```
g.drawRect(x, y, width, height);
```

The *drawRect* method is currently implemented with a series of four draw lines. We hope this will change as the AWT becomes more optimized.

To clear a rectangular area of the screen, the *Graphics* class draws a rectangle with the current background color:

```
g.clearRect(x, y, width, height);
```

The *Graphics* class provides rounded shape drawing. When a rounded shape is drawn, *arcWidth* and *arcHeight* are used. *arcWidth* and *arcHeight* are the width and height of an ellipse used to draw the rounded corners.

Here's how to draw the outline of a rectangle with rounded corners using the current color:

```
g.drawRoundRect(x, y, width, height, arcWidth, arcHeight);
```

To draw a filled rectangle with rounded corners using the current color:

```
g.fillRoundRect(x, y, width, height, arcWidth, arcHeight);
```

To draw a highlighted 3-D rectangle using two colors:

```
g.draw3DRect(x, y, width, height, raised);
```

If *raised* is true, the rectangle appears raised. This effect is currently implemented using darker and brighter colors to create highlights. The same rectangle can also be drawn filled:

```
g.fill3DRect(x, y, width, height, raised);
```

To draw an oval outline or a filled oval, use this code:

```
g.drawOval(x, y, width, height);
g.fillOval(x, y, width, height);
```

To draw an arc inscribed in a rectangle, starting at *startAngle* and ending at *startAngle+arcAngle*, use the following:

```
g.drawArc(x, y, width, height, startAngle, arcAngle);
```

Angles are in degrees. Zero degrees is parallel to the X-axis, and positive angles are measured in a counterclockwise direction. For the filled form of the arc:

```
g.fillArc(x, y, width, height, startAngle, arcAngle);
```

To draw a polygon using an array of (x,y) coordinates:

```
g.drawPolygon(xArray, yArray, nPoints);
```

To draw a polygon using an instance of the *Polygon* class:

```
g.drawPolygon(aPolygon);
```

The *drawPolygon* method is implemented with a call to

```
drawPolygon(aPolygon.xpoints, aPolygon.ypoints, aPolygon.npoints);
```

To fill the polygon with the current color:

```
g.fillPolygon(xArray, yArray, nPoints);
```

There must be more than three points or else the polygon call is ignored. If the polygon has overlapping parts, the even-odd fill rule is used. This is typically implemented using a scanline that is drawn from left to right. If the scanline crosses the edge of the polygon an odd number of times, the pixels are inside the polygon and will be filled; otherwise, they fall outside the polygon. This is a typical approach for filling concave polygons.

```
g.fillPolygon(aPolygon);
```

To draw a string starting at (x,y) using the current font and color:

```
String str = "Java is fun";
g.drawString(str,x,y);
```

To draw a character or byte array starting at (x,y) using the current font and color:

```
g.drawChars(charArray, offsetIntoArray, numberOfItemToDraw, x, y);
g.drawBytes(byteArray,  offsetIntoArray, numberOfCharactersToDraw, x, y);
```

The *drawBytes* method assumes that the bytes represent the least significant eight bits of the Unicode characters. To draw an image at (x,y) and notify *anImageObserver*:

```
aBoolean =  g.drawImage(img, x, y, anImageObserver);
```

To draw an image inside a rectangle (with optional scaling):

```
aBoolean =  g.drawImage(img, x, y, width, height, anImageObserver);
```

To draw an image with a background color:

```
aBoolean =  g.drawImage(img, x, y, aColor, anImageObserver);
```

To draw an image with a background color and in a rectangle:

```
aBoolean =  g.drawImage(img, x, y, width, height, aColor, anImageObserver);
```

To dispose of the graphics context (before the garbage collector does):

```
g.dispose();
```

The *Graphics* class overrides *Object.finalize()* with a dispose call:

```
g.finalize();
```

To convert the *Graphics* instance into a string:

```
g.toString();
```

The following sections give practical examples for the use of the methods in the *Graphics* class.

How to Draw a Grid

In this section you will learn how to write a class that can draw a grid into a *Graphics* instance. Consider the following code for drawing a grid:

```
1.      import java.applet.*;
2.      import java.util.*;
3.      import java.awt.*;
```

Lines 1–3 are used to import packages into your program. Line 4 declares the class as a *final* class, preventing other programmers from subclassing the *draw* class.

```
4.      final public class draw {
5.      // prevent instantiation
6.      private draw() {}
```

Line 6 declares the constructor for the *draw* class to be *private*. This prevents the *draw* class from being instanced by programmers who develop outside the *draw* class. Line 7 is the *grid* method, which takes an instance of a *Graphics* class. The dimensions of the clip rectangle are extracted from a *Rectangle* instance using the *width* and *height* members of the *Rectangle* class. The *width* and *height* are stored as integers in units of pixels.

```
7.      static void  grid(Graphics g) {
8.      Rectangle r = g.getClipRect();
```

The *Geometry* class is a custom class used by DiffCAD to store global instances. *Grid_spacing* is an *int* whose units are pixels.

```
9.     int grid_spacing = Geometry.grid_width.getValue();
10.     int w = r.width;
```

```
    int h = r.height;
    for (int x = 0; x < w; x = x + grid_spacing)
            {g.drawLine(x,0,x,h);}
       for (int y = 0; y < h; y = y + grid_spacing)
        {g.drawLine(0,y,w,y);}
  }
    }
```

The FontMetrics Class

The *FontMetrics* class is a public abstract class in the *java.awt* package. Font metrics are based on dimensions that are special to fonts. Font dimensions include the baseline, ascent, descent, leading, and height. Distances are generally given in pixels (when they are returned by the *FontMetrics* attribute methods). The font metrics dimensions are shown in Figure 3.2

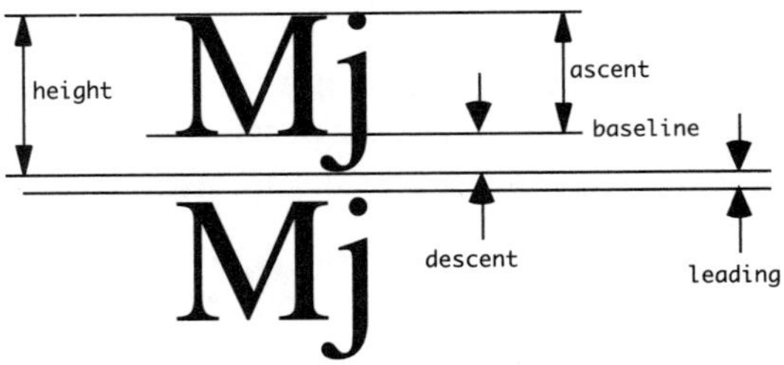

FIGURE 3.2 FONT METRICS DIMENSIONS.

Lines of cast type, called *slugs*, were originally set by hand; the compositor inserted a lead strip between slugs. That's why the blank space between lines of type is called *leading*. The Linotype, the first practical typesetting machine, was used in 1886 at the *New York Tribune* [Mertle]. The nomenclature of typesetting has passed to the electronic era essentially unchanged. Typical font metrics are in units of points (at 72 points per inch). Because of the raster orientation of the *java.awt* , all the font metrics are measured in pixels. The number of pixels per inch is called the display's *pitch*. Thus, there is no display–independent way to automatically relate the font metrics of the AWT to the font metrics of traditional typesetting. Such a system would have to be able to detect the pitch of the display before the traditional font metrics could be computed.

Class Summary

```
package java.awt;
public abstract class FontMetrics {
    public Font getFont()
    public int getLeading()
    public int getAscent()
    public int getDescent()
    public int getHeight()
    public int getMaxAscent()
    public int getMaxDescent()
    public int getMaxDecent()
    public int getMaxAdvance()
    public int charWidth(int ch)
    public int charWidth(char ch)
    public int stringWidth(String str)
    public int charsWidth(char data[], int off, int len)
    public int bytesWidth(byte data[], int off, int len)
    public int[] getWidths()
    public String toString()
}
```

Class Usage

Suppose that the following constants are defined:

```
Graphics g;
FontMetrics theFontMetrics = g.getFontMetrics();
Font aFont;
int distanceInPixels;
char ch;
char charArray[];
byte byteArray[];
int offset, length
```

Here's how to get the instance of the *Font* class upon which the *FontMetrics* are based:

```
afont = theFontMetrics.getFont();
```

To get the standard leading (line spacing between descent and ascent):

```
distanceInPixels = theFontMetrics.getLeading();
```

To get the ascent, descent, and height:

```
distanceInPixels = theFontMetrics.getAscent();
distanceInPixels = theFontMetrics.getDescent();
distanceInPixels = theFontMetrics.getHeight();
```

To get the maximum ascent and descent of a font:

```
distanceInPixels = theFontMetrics.getMaxAscent();
distanceInPixels = theFontMetrics.getMaxDescent();
```

To get the maximum height of a font, add the maximum descent and ascent. To get the width of a character in the font, use this code:

```
distanceInPixels = theFontMetrics.charWidth(ch);
```

To get the width of a string in the font:

```
distanceInPixels = theFontMetrics.stringWidth(str);
```

To get the width of an array of characters in the font, starting at the offset and proceeding for *length* characters:

```
distanceInPixels = theFontMetrics.charsWidth(charArray, offset, length);
```

To get the width of an array of bytes in the font, starting at the offset and proceeding for *length* characters (characters come in as bytes and are converted to 16–bit characters):

```
distanceInPixels = theFontMetrics.bytesWidth(byteArray, offset, length);
```

To get the width of the first 256 characters in the font:

```
distanceInPixels = theFontMetrics.getWidths();
```

To get the string representation of *FontMetrics*:

```
str = theFontMetrics.toString();
```

How to Draw a String with a Background

Often the user will wish to draw a string that has a superimposed background. This is important if the string is to be changed dynamically. In the following example we show how to draw the date and time. The *clearRect* call erases a part of the display so that the string does not overwrite itself.

```
synchronized private void draw() {
Dimension dim = f.size();
int height = dim.height - 60;
int width = dim.width;
Date theDate = new Date();
String date_string = theDate.toString();
int xloc = 10;
int yloc = dim.height - 60;
int string_width = getFontMetrics(
          g.getFont()).stringWidth(date_string);
int string_height = getFontMetrics(
          g.getFont()).getHeight();
g.clearRect(xloc,yloc,string_width,string_height);
g.drawString(date_string, xloc,height+xloc);
}
```

An example of the string, with filled background in place, is shown in Figure 3.3.

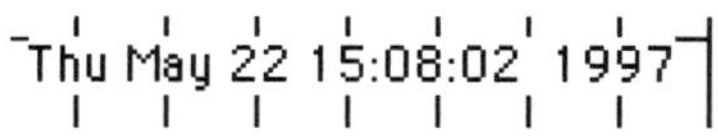

FIGURE 3.3 STRING WITH FILLED BACKGROUND.

How to Draw a Vertical String

The following method draws a string vertically (something the AWT does not normally do):

```
public void drawVerticalString(Graphics g,String str,int x,int y) {
int str_height = g.getFontMetrics().getHeight();
(str_height*str.length())/2;
for (int i = 0; i<str.length(); i++) {
     int char_width = g.getFontMetrics().stringWidth(str.substring(i,i+1));
     g.drawString(str.substring(i,i+1),x-char_width/2,y);
     y+=str_height;
} // end for
} // end drawVerticalString
```

An example of the use of the vertical string is shown in Figure 3.4.

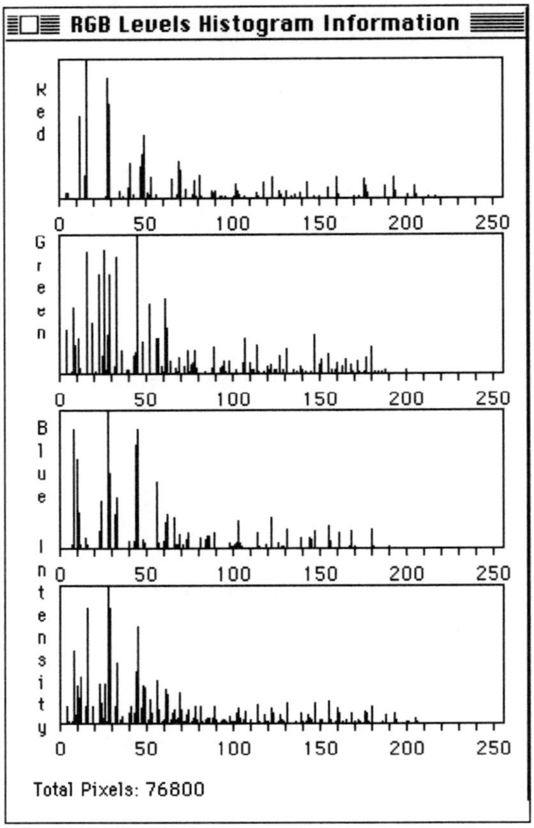

FIGURE 3.4 AN EXAMPLE OF THE VERTICAL STRING: THE HISTOGRAM.

The MenuItem Class

An instance of the *MenuItem* class is typically added to a *Menu* instance. When an event is posted by a component, an instance of the *MenuItem* class can be passed in the *target* member of the *Event* instance. Thus, it is typical to write *handleEvent* methods that compare targets with *MenuItem* instances to see what the user has selected. The event handler routines are discussed in more detail later in this chapter.

Class Summary

```
public class MenuItem extends java.awt.MenuComponent {
/* Instance Variables */
    boolean enabled;
    String label;
/* Methods */
    public void MenuItem(String label);
    public synchronized void addNotify();
    public String getLabel();
    public void setLabel(String label);
    public boolean isEnabled();
    public void enable();
    public void enable(boolean on);
    public void disable();
    public String paramString();
}
```

Class Usage

The *MenuItem* class is a subclass of *MenuComponent* and is used to store a string item that represents a menu choice. Assume that the following variables are predefined:

```
MenuItem ifft_mi          = new MenuItem("[2] IFFT");
String label;
boolean on;
```

To get and set the string contained in the *MenuItem* instance, use the following:

```
str = ifft_mi.getLabel();
ifft_mi.setLabel(label);
```

To get and set the enabling on the *MenuItem* instance:

```
on = ifft_mi.isEnabled();
ifft_mi.enable(); // always enables
ifft_mi.enable(on); // conditionally enables
ifft_mi.disable(); // always disables
```

To obtain *"label="* + *ifft_mi.getLabel()*, use this:

```
label = ifft_mi.paramString();
```

NOTE

The *MenuItem* class has a special "-" label that is used as a separator. An example of the separator's appearance in the menu is shown in Figure 3.5.

FIGURE 3.5 THE SEPARATOR IS SHOWN BETWEEN SENDMAIL AND LS.

The Event Class

The *Event* class is a reference data type that is a direct descendent of the *java.lang.Object* class. The *handleEvent* method is invoked with an instance of the *Event* class whenever user input is detected. *Components* are said to *post* events.

Class Summary

```
public class Event {
    public static final int SHIFT_MASK
    public static final int CTRL_MASK
    public static final int META_MASK
    public static final int ALT_MASK
    public static final int HOME
    public static final int END
    public static final int PGUP
    public static final int PGDN
    public static final int UP
```

```
public static final int DOWN
public static final int LEFT
public static final int RIGHT
public static final int F1
public static final int F2
public static final int F3
public static final int F4
public static final int F5
public static final int F6
public static final int F7
public static final int F8
public static final int F9
public static final int F10
public static final int F11
public static final int F12
public static final int WINDOW_DESTROY
public static final int WINDOW_EXPOSE
public static final int WINDOW_ICONIFY
public static final int WINDOW_DEICONIFY
public static final int WINDOW_MOVED
public static final int KEY_PRESS
public static final int KEY_RELEASE
public static final int KEY_ACTION
public static final int KEY_ACTION_RELEASE
public static final int MOUSE_DOWN
public static final int MOUSE_UP
public static final int MOUSE_MOVE
public static final int MOUSE_ENTER
public static final int MOUSE_EXIT
public static final int MOUSE_DRAG
public static final int SCROLL_LINE_UP
public static final int SCROLL_LINE_DOWN
public static final int SCROLL_PAGE_UP
public static final int SCROLL_PAGE_DOWN
public static final int SCROLL_ABSOLUTE
public static final int LIST_SELECT
public static final int LIST_DESELECT
public static final int ACTION_EVENT
public static final int LOAD_FILE
public static final int SAVE_FILE
public static final int GOT_FOCUS
public static final int LOST_FOCUS
public Object target;
public long when;
public int id;
public int x;
public int y;
public int key;
public int modifiers;
public int clickCount;
public Object arg;
```

```
    public Event evt;
/* Methods */
    public Event(Object target, long when, int id, int x, int y, int key,int modi-
fiers, Object arg)
    public Event(Object target, long when, int id, int x, int y, int key, int modifiers)
    public Event(Object target, int id, Object arg)
    public void translate(int x, int y)
    public boolean shiftDown()
    public boolean controlDown()
    public boolean metaDown()
    protected String paramString()
    public String toString()
}
```

Class Usage

Suppose the following constants are predefined:

```
Object target;
long when; // a creation time stamp
int id; // the event type
int x, y; // event location
int key;
int modifiers;
Object arg;
```

The *target* is the instance of the *Component* class that posted the event. The *when* member holds a creation time stamp measured in milliseconds. The *id* permits decoding of the event type with the use of a *switch* statement. The location of the event is contained in (x,y). The state of the modifier keys is encoded into the *modifiers* parameter. *Arg* is the instance of the class that is to be associated with the event. The *Event* class constructor is overloaded so that we can write the following:

```
Event evt = new Event(target, id, arg);
Event evt = new Event(target, when, id, x, y, key,      modifiers);
Event evt = new Event(target, when, id, x, y, key,      modifiers, arg);
```

Typically, it is the user's job to write code to decode the events. Unfortunately, the services provided by the AWT lead to an ad hoc approach for event handling that results in poor software engineering. The AWT assumes that keyboard events and menu choice events should be handled in different parts of the source code. Often, keyboard shortcuts are used to speed menu selections. Thus, menu selections are processed in a different part of the source code from the keyboard event, leading to duplication of code. Additional development effort is needed to add keyboard shortcuts. A remedy for this design flaw in the AWT is discussed in the following section.

Several public variables are associated with the *Event* class. Some of the public variables are key masks designed to help with decoding. An example of this is shown in the *Component* section:

```
KEY_EVENT, SHIFT_MASK, CTRL_MASK,META_MASK, ALT_MASK, HOME, END, PGUP,
PGDN,KEY_PRESS, KEY_RELEASE, KEY_ACTION, KEY_ACTION_RELEASE
```

The arrow keys are decoded with the following:

```
UP, DOWN, LEFT, RIGHT
```

The function keys are decoded with these variables:

```
F1, F2, F3, F4, F5, F6, F7, F8, F9, F10, F11, F12
```

The window events are decoded with

```
WINDOW_EVENT, WINDOW_DESTROY, WINDOW_EXPOSE, WINDOW_ICONIFY, WINDOW_DEICONIFY,
WINDOW_MOVED, WINDOW_EVENT
```

The mouse events are decoded with

```
MOUSE_EVENT, MOUSE_DOWN, MOUSE_UP, MOUSE_MOVE, MOUSE_ENTER, MOUSE_EXIT, MOUSE_DRAG
```

The scrolling events are decoded with

```
SCROLL_EVENT, SCROLL_LINE_UP, SCROLL_LINE_DOWN, SCROLL_PAGE_UP, SCROLL_PAGE_DOWN,
SCROLL_ABSOLUTE
```

The list events are decoded with

```
LIST_EVENT, LIST_SELECT, LIST_DESELECT
```

A series of events are classified as miscellaneous events:

```
MISC_EVENT, ACTION_EVENT,LOAD_FILE, SAVE_FILE, GOT_FOCUS, LOST_FOCUS
```

Event Handling

In part because of its modularization, the AWT model for handling events can lead to poor software engineering. The AWT designers sought to map mouse events, keyboard events, and menu selection events using a preprocessing facility built into the *Component* class. This policy is based on the assumption that the relationship between event production and code invocation is isomorphic. This assumption is fallacious because multiple events can often be used to trigger a single task. Menu choices often have keyboard shortcuts, and both a menu event and a keypress event can trigger the same processing.

The Keyboard

To handle keyboard events in the AWT, a class that extends the *Component* class will typically have its *keyDown* method invoked. In the following code fragment, the *AudioFrame* class extends the *Frame* class (which is two generations removed from the *Component* superclass):

```
...
public class AudioFrame extends Frame {
...
public boolean keyDown(Event e, int key) {
     switch (key) {
             case '2':
                  ifft();
                  return true;
        }
        System.out.println("Unknown key function:"+ key);
        return true;
}
```

The Target

To handle the passing of an instance, the *Event* class supports a *target* member that is of *Object* type. To process this event, a *handleEvent* method is used:

```
...
public class AudioFrame extends Frame {
...
Menu m = new Menu("Audio Menu");
...
MenuItem ifft_mi = addItem("[2] IFFT");
     public MenuItem addItem(String itemName) {
          MenuItem mi = new MenuItem(itemName);
          m.add(mi);
          return(mi);
     }
     public void init_menu() {
          // my menu items
          MenuBar menuBar = new MenuBar();
```

```
          // Initialize the menu bar
          setMenuBar(menuBar);
         menuBar.add(m);
...
public boolean handleEvent(Event e) {
...
          if (e.target == ifft_mi) {
               ifft();
               return true;
          }
...
```

Notice that the keyboard event calls the same code that the *handleEvent* calls, because the *MenuItem* instance (which appears in the main menu bar) has a keyboard shortcut. This is unsound software engineering, because parallel maintenance must be performed in both the keyboard and menu event processing code. A sounder approach appears in the following section.

The Evt Class

The *Evt* class is a public static class that contains a series of methods that address the software engineering problem cited in the preceding section. The basic assumption is that several different event *combinations* can cause a single result, the example being a keypress event and a menu item event. The solution involves the creation of an overloaded *match* method:

```
class Evt   {
...
  public static boolean matchKey(Event e, int key) {
       return ((e.id == Event.KEY_PRESS) && (e.key==key));
  }
  public static  boolean match(Event e, int key, String str) {
       return (matchKey(e,key) || e.arg.equals(str));
  }

  public static  boolean match(Event e, int key, Object target) {
       return (matchKey(e,key) || e.target.equals(target));
  }
...
```

The *Evt* class offers the immediate benefit of giving a two-way match between keyboard and *menuItem* events. In the *AudioFrame* class, the *keyDown* method is eliminated and replaced with this:

```
public boolean handleEvent(Event e) {
...
          if (Evt.match(e,'2',ifft_mi)) {
```

```
            ifft();
            return true;
        }
...
```


This single convention eliminates 60 lines of code in the *AudioFrame keyDown* event handler. To add another level of automation to the event processing, we assume that the keyboard shortcut will be encoded into the string representation of the *menuItem*, as it appears in the main menu bar. Figure 3.6 shows a sample main menu bar from the *AudioFrame* in the DiffCAD program.

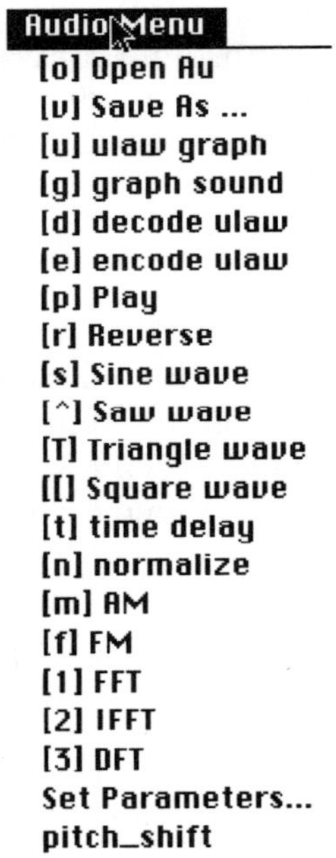

FIGURE 3.6 SAMPLE AUDIOFRAME MAIN MENU BAR DROP-DOWN MENU.

One feature of the *AudioFrame* keyboard shortcuts is that they conform to this convention: If a keyboard shortcut exists, it is encoded by a [followed by a single character. Tognazzini reports that $50 million of research led to two conclusions: Users say keyboard shortcuts are faster, and the stopwatch shows that mouse choices are faster. Tognazzini produced a guideline that states that visual interface construction should not have its resources sapped by the keyboard interface [Tognazzini]. As a result of this common sense suggestion, we have combined the following check into a single method called *match*:

```
public static  boolean match(Event e, Object target) {
     int c = getKeyboardShortCut(e, target);
 return match(e, c, target);
}
```

In the following code, *getKeyboardShortCut* makes three assumptions if a keyboard shortcut is embedded in the target: The target will be an instance of a *MenuItem*, the first character of the label will be a [and the second character of the label will be the keyboard shortcut character.

```
public static int getKeybcardShortCut(Event e, Object target) {
     if (target instanceof MenuItem) {
          MenuItem mi = (MenuItem) target;
          String str = mi.getLabel();
          int index = str.indexOf('[');
          if (index == 0) { return str.charAt(1);}
          }
     return -1;
}
```

This approach leads to a more consistent interface (keyboard shortcuts are always encoded visually and consistently), an easier–to–use API (only one cal, instead of two), and better software engineering (centralized interface changes with fewer lines of code). Keyboard and menu item events are now handled as follows:

```
...
public boolean handleEvent(Event e) {
...
          if (Evt.match(e,ifft_mi)) {
               ifft();
               return true;
          }
...
```

It takes no more effort to encode and process a keyboard shortcut than to process and set up a visual interface, fulfilling Tognazzini's basic guideline.:

```
     MenuItem ifft_mi          = addItem("[2] IFFT");
```

The Mouse

Any pick device (such as a trackball, penlight, touch pad, or tablet) can generate mouse events. The Java AWT supports only a single pick device, and, no matter what the pick device is, it is said to generate mouse events. To handle the mouse events, the AWT *Component* class provides a suite of callback methods.

When the mouse button is depressed:

```
public boolean mouseDown(Event evt, int x, int y)
```

When the mouse button is depressed and the mouse is moved:

```
public boolean mouseDrag(Event evt, int x, int y)
```

When the mouse button is released:

```
public boolean mouseUp(Event evt, int x, int y)
```

When the mouse is moved:

```
public boolean mouseMove(Event evt, int x, int y)
```

When the mouse is over a component:

```
public boolean mouseEnter(Event evt, int x, int y)
```

When the mouse leaves the component:

```
public boolean mouseExit(Event evt, int x, int y)
```

To implement a mouse event, you must either override the preceding events or use a *switch* statement to process the *Event.id*. The choice between *switch* statement and method is a question of preference, policy, and good software engineering practice. The *switch* statement is more flexible and permits more "combination" events (such as a shift-click for extending a selection).

CD-ROM

In the following code fragment taken from *GUI.java*, a *handleEvent* method is implemented with a *switch* statement

```
1. public boolean handleEvent(Event e) {
2.     switch (e.id) {
```

On line 3, the *Event.MOUSE_UP* indicates that the mouse button was released. When this occurs, the *processMouseUp* method is invoked, causing an object to change its appearance.

```
3.         case Event.MOUSE_UP: {
4.         String objectName = mouse_choice.getSelectedItem();
5.         processMouseUp( objectName, e.x, e.y);
6.         repaint();
7.         }
```

On line 8, the *Event.MOUSE_DOWN* triggers the recording of the location of the mouse cursor. The anchor point is used to provide relative mouse motion.

```
8.         case Event.MOUSE_DOWN:
9.              anchor = new point( e.x, e.y);


10.    } // switch
```

The Component Class

The *Component* class is an abstract class. Subclasses of components are used to make instances that can be displayed on the screen. Examples of such component subclasses include *Button*, *Canvas*, *Checkbox*, *Choice*, *Container*, *Label*, *List*, *Scrollbar*, and *TextComponent*. Figure 3.7 shows the *Component* hierarchy

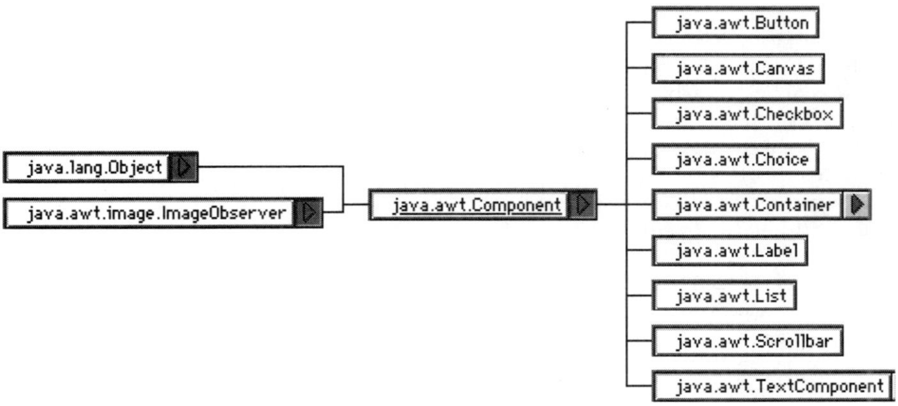

FIGURE 3.7 COMPONENT HIERARCHY.

The *Component* class resides in the *java.awt* package and implements the *ImageObserver* interface. A *Component* instance supports services that include drawing, event handling, font control, color control, image handling, and size and position control. These services are provided by a series of methods and instance variables.

A suite of methods is available that supports event handling: *handleEvent, mouseEnter, mouseExit, mouseMove, mouseDown, mouseDrag, mouseUp, keyDown,* and *action.* The *Component* class provides a mechanism for invoking the specialized event handlers. This mechanism is based on a *switch* statement that calls methods that are to be implemented by subclasses. The methods are implemented in the *Component.java* class by returning false, signaling that the event was not processed.

NOTE

Some components are used for input, others for output, and still others for both input and output. In a later section we will see an example of a button component that is used only as an input. We will also see an example of a label component, that serves only as an output and a scrollbar component that can serve as both an input and an outputcComponent.

Class Summary

```
public abstract class Component implements ImageObserver {
public Container getParent()
public ComponentPeer getPeer()
public Toolkit getToolkit()
public boolean isValid()
public boolean isVisible()
public boolean isShowing()
public boolean isEnabled()
public Point location()
public Dimension size()
public Rectangle bounds()
public synchronized void enable()
public void enable(boolean cond)
public synchronized void disable()
public synchronized void show()
public void show(boolean cond)
public synchronized void hide()
public Color getForeground()
public synchronized void setForeground(Color c)
public Color getBackground()
public synchronized void setBackground(Color c)
public Font getFont()
public synchronized void setFont(Font f)
public synchronized ColorModel getColorModel()
public void move(int x, int y)
public void resize(int width, int height)
public void resize(Dimension d)
```

```
public synchronized void reshape(int x, int y, int width, int height)
public Dimension preferredSize()
public Dimension minimumSize()
public void layout()
public void validate()
public void invalidate()
public Graphics getGraphics()
public FontMetrics getFontMetrics(Font font)
public void paint(Graphics g)
public void update(Graphics g)
public void paintAll(Graphics g)
public void repaint()
public void repaint(long tm)
public void repaint(int x, int y, int width, int height)
public void repaint(long tm, int x, int y, int width, int height)
public void print(Graphics g)
public void printAll(Graphics g)
public boolean imageUpdate(Image img, int flags,int x, int y, int w, int h)
public Image createImage(ImageProducer producer)
public Image createImage(int width, int height)
public boolean prepareImage(Image image, ImageObserver observer)
public boolean prepareImage(Image image, int width, int height, ImageObserver
observer)
public int checkImage(Image image, ImageObserver observer)
public int checkImage(Image image, int width, int height,ImageObserver observer)
public synchronized boolean inside(int x, int y)
public Component locate(int x, int y)
public void deliverEvent(Event e)
public boolean postEvent(Event e)
public boolean handleEvent(Event evt)
public boolean mouseDown(Event evt, int x, int y)
public boolean mouseDrag(Event evt, int x, int y)
public boolean mouseUp(Event evt, int x, int y)
public boolean mouseMove(Event evt, int x, int y)
public boolean mouseEnter(Event evt, int x, int y)
public boolean mouseExit(Event evt, int x, int y)
public boolean keyDown(Event evt, int key)
public boolean keyUp(Event evt, int key)
public boolean action(Event evt, Object what)
public void addNotify()
public synchronized void removeNotify()
public boolean gotFocus(Event evt, Object what)
public boolean lostFocus(Event evt, Object what)
public void requestFocus()
public void nextFocus()
public String toString()
public void list()
public void list(PrintStream out)
public void list(PrintStream out, int indent)
}
```

Class Usage

We use the term *peer* to indicate the parallel that exists in the native implementation of a component. For example, *FileDialog* is a subclass of a *Component* that has a different implementation on every platform. To provide an API that looks the same on all platforms, Java uses a peer class that provides the services needed to allow the API to function properly. This means that the Open dialog box created on a Macintosh is the standard File Open dialog box that is supplied by the operating system. Also, the Open dialog box on a Windows system is the Load File dialog box that is native to Windows. Peers are the mechanism by which Java achieves a portable operating system interface.

Suppose the following variables are defined:

```
Container parent;
int x,y;
int width, height;
Color foreground, background;
Font font;
boolean visible = true;
boolean enabled = true;
boolean valid = false;
Component component;
ComponentPeer componentPeer;
Dimension aDimension;
Graphics graphics;
Image anImage;
Boolean aBoolean;
int flags, indent;
ImageProducer anImageProducer;
```

To get the parent of the component, use this code:

```
parent = component.getParent();
```

To get the peer of the component:

```
componentPeer = component.getPeer();
```

To get the toolkit used by the component (see the *Toolkit* class):

```
toolKit = component.getToolkit();
```

To see whether the peer is known to the component and whether the component is properly laid out:

```
valid = component.isValid();
```

To see whether the Component is visible (the *hide* method can make this statement return false):

```
visible = component.isVisible();
```

Visible components may not be showing:

```
visible = component.isShowing();
```

Enabled components can generate events:

```
aBoolean = isEnabled();
```

To unconditionally enable a component:

```
component.enable();
```

To conditionally enable a component:

```
component.enable(cond);
```

To disable a component:

```
component.disable();
```

To get the location of the component relative to the parent's coordinate system:

```
aPoint = component.location();
```

To get the dimensions and bounds of the component:

```
aDimension = component.size();
aRectangle = component.bounds();
```

To show the component:

```
component.show();
```

To conditionally show the component:

```
component.show(cond);
```

To hide the component:

```
component.hide();
```

To get the component's foreground color (or that of its parent):

```
aColor = component.getForeground();
```

To set the component's foreground color:

```
component.setForeground(aColor);
```

To get the component's background color (or that of its parent):

```
aColor = component.getBackground();
```

To set the component's background color:

```
component.setBackground(aColor);
```

To get the font from a component or that of its parent if the component does not have one set:

```
font = component.getFont();
```

To set the font of a component:

```
component.setFont(font);
```

To get the *ColorModel* used for display:

```
colorModel = component.getColorModel();
```

To move the component using the parent's coordinate system:

```
component.move(x, y);
```

To resize the component:

```
component.resize(width, height);
component.resize(aDimension);
```

To reshape (resize and move) a component, in the parent's coordinates:

```
component.reshape(x, y, width, height);
```

To get the preferred or minimum dimensions for the component:

```
aDimension = component.preferredSize();
aDimension = component.minimumSize();
```

To lay out the component (typically, *validate* is used):

```
component.layout();
```

To lay out the component and make the component valid:

```
component.validate();
```

To invalidate a component and its parent (marking for layout):

```
component.invalidate();
```

To get a *Graphics* context for the component (and to return *null* if there is no peer):

```
component.getGraphics();
```

To get the font metrics for the component's peer, if available, otherwise returning the font metrics for the component:

```
component.getFontMetrics(font);
```

To paint the component:

```
component.paint(graphics);
```

To update the component (typically called by *repaint*):

```
component.update(graphics);
```

To paint the component and the subcomponents:

```
component.paintAll(graphics);
```

To schedule a call to update:

```
component.repaint();
```

To schedule a call to update for a specific interval:

```
component.repaint(milliseconds);
component.repaint(x, y, width, height);
```

To schedule a call to update for a part of the component in a specific time:

```
component.repaint(milliseconds, x, y, width, height);
```

To paint the component on the *graphics* context (typically overridden):

```
component.print(graphics);
```

To invoke *print* on the component and subcomponents:

```
component.printAll(graphics);
```

To repaint the component after the image has been delivered (typically called by an image observer and returns *false* if image has not changed):

```
aBoolean = component.imageUpdate(anImage, flags, x, y, width, height);
```

To create an image from the image producer (see *ImageProducer* for more information):

```
anImage = component.createImage(anImageProducer);
```

To create an off-screen image for double buffering:

```
anImage = component.createImage(width, height);
```

To prepare an image for display on the component (this starts a thread; see the *ImageObserver*):

```
component.prepareImage(anImage, anImageObservser);
```

To prepare the image with a particular width and height:

```
component.prepareImage(anImage, width, height, anImageObservser);
```

To obtain the status of the screen representation of an image:

```
statusInt = component.checkImage(anImage, anImageObserver);
```

See *ImageObserver* to decode *statusInt*.
To get the status for an image to be scaled:

```
statusInt = component.checkImage(anImage, width, height, anImageObserver);
```

To see whether a location is inside a component's borders relative to the component's coordinate system:

```
aBoolean = component.inside(x,y);
```

To get the component containing a location:

```
component = locate( x, y);
```

To post an event:

```
component.deliverEvent(anEvent); // the same as postEvent(anEvent)
```

To call *handleEvent* with an event on a component (or, if the component does not handle it, the parent of the component):

```
component.postEvent(anEvent);
```

See the discussion in the preceding section for a description of the events and how to handle them.

The component provides call backmethods:

```
public boolean mouseDown(Event evt, int x, int y)
public boolean mouseDrag(Event evt, int x, int y)
public boolean mouseUp(Event evt, int x, int y)
public boolean mouseMove(Event evt, int x, int y)
public boolean mouseEnter(Event evt, int x, int y)
public boolean mouseExit(Event evt, int x, int y)
public boolean keyDown(Event evt, int key)
public boolean keyUp(Event evt, int key)
public boolean action(Event evt, Object what)
public boolean gotFocus(Event evt, Object what)
public boolean lostFocus(Event evt, Object what)
```

An invocation of *show* or *pack* will call *addNotify* (the programmer typically does not make this call). *AddNotify* makes the component invalid and causes a peer to be created:

```
component.addNotify();
```

To dispose of a components peer:

```
component.removeNotify();
```

To get the keyboard focus (i.e., keystrokes stream into your component):

```
component.requestFocus();
```

To advance the keyboard focus to the next component:

```
component.nextFocus();
```

To make a string representation of the component:

```
component.toString();
```

To invoke the *list* method on *System.out*:

```
component.list();
```

To invoke the *list* method on a *PrintStream*:

```
component.list(aPrintStream);
```

To call *toString* and print the component and all its children, with indentation:

```
component.list(aPrintStream, indent);
```

The Container Class

The *Container* class is an extension of the *Component* class that can hold *Component* instances. The instances that are added to an instance of a *Container* class are called *children*. The *Container* class that holds the children is called the *parent*. Because a container can hold other containers, a long line of descendants is possible. A parent can arrange the appearance of the children using a layout manager. Each child has its own layout manager. If a container (or its children) is not properly laid out, the layout is said to be invalid. A parent lays out the children and all their descendants during the layout process. Containers that are valid are not laid out needlessly.

The *Container* class structure is that of an AKO (a-kind-of) taxonomy. This is different from the *Container* hierarchy, which is based on has-a relationships. For example, a *Frame* has-a *Panel* that has-a *Checkbox*. Both a *Frame* and a *Panel* are kinds-of containers. Containers always have an *add* method that provides the implementation for adding either components or other containers to the has-a hierarchy. A *Container* instance can handle events with its own *handleEvent* implementation. Recall that *getParent* is inherited by any subclass of *Component* (such as the *Container* class).

Class Summary

```
public abstract class Container extends Component
   public int countComponents()
   public synchronized Component getComponent(int n)
   public synchronized Component[] getComponents()
   public Insets insets()
   public Component add(Component comp)
```

```
    public synchronized Component add(Component comp, int pos)
    public synchronized Component add(String name, Component comp)
    public synchronized void remove(Component comp)
    public synchronized void removeAll()
    public LayoutManager getLayout()
    public void setLayout(LayoutManager mgr)
    public synchronized void layout()
    public synchronized void validate()
    public synchronized Dimension preferredSize()
    public synchronized Dimension minimumSize()
    public void paintComponents(Graphics g)
    public void printComponents(Graphics g)
    public void deliverEvent(Event e)
    public Component locate(int x, int y)
    public synchronized void addNotify()
    public synchronized void removeNotify()
    public void list(PrintStream out, int indent)
}
```

Class Usage

Suppose that the following variables are defined:

```
Component comp, compArray[];
Container cont;
Insets insets;
int ncomponents, i;
String layoutName; // the name of a layout manager.
LayoutManager layMan;
Graphics gc;
int x,y;
PrintStream printStream;
int indent;
```

To get the number of components use this statement:

```
ncomponents = comp.countComponents();
```

To get the *n*th component:

```
comp = cont.getComponent(i);
```

To get all the components:

```
compArray = cont.getComponents();
```

To get the insets (rectangular dimensions in pixels from the edges):

```
insets = cont.insets();
i = insets.bottom;
i = insets.top;
i = insets.left;
i = insets.right;
```

To add a *Component* to a *Container*:

```
cont.add(comp);
```

To add a *Component* to a *Container* in a given order (–1 means end of list):

```
cont.add(comp, i);
```

To add a *Component* to a *Container* using a layout manager:

```
cont.add(layoutName, i);
```

To remove a *Component* from a *Container*:

```
cont.remove(comp);
```

To delete all *Components*:

```
cont.removeAll();
```

To get and set the layout manager:

```
layMan = cont.getLayout();
cont.setLayout(layMan);
```

To perform layout on the *Container* but not the descendants:

```
cont.layout();
```

To perform layout on the *Container* and all descendants:

```
cont.validate();
```

To get the preferred and minimum dimensions:

```
Dimension dim = cont.preferredSize();
dim cont.minimumSize();
```

To paint or print the *Components* onto a *Graphics* instance:

```
cont.paintComponents(gc);
cont.printComponents(gc);
```

From within a *Container* (such as a *Frame*) you can write the following:

```
printComponents(getGraphics());
paintComponents(getGraphics());
```

To post an *Event* to a *Component* located at (e.x, e.y):

```
Event e;
cont.deliverEvent(e);
```

To locate a component at (x, y):

```
comp = cont.locate(x, y); // null returned if compoent unfound
```

To send the *addNotify* message to all descendents and the superclass:

```
cont.addNotify(); // this creates peer descendents
```

To send the *removeNotify* message to all descendents and the superclass:

```
cont.removeNotify(); // this deletes peer descendents
```

To print a list of the descendents:

```
cont.list(printStream, indent);
```

The *indent* is the indentation used during generation traversal. For example, in *AudioFrame.java*, a *MenuItem* event is handled by the following:

```
if (Evt.match(e,print_mi)) {
          validate();
          list(System.out,5);
          return true;
}
```

The preceding code emits a component list, including four Scroll bars, a panel, a label, and an *intLabel*:

```
lyon.AudioFrame[0,0,550x385,layout=java.awt.BorderLayout,title=/hd/current/Java%20
book/code/au/ah.au]
```

```
java.awt.Scrollbar[0,0,550x16,val=14,vis=true,min=0,max=24,horz]
java.awt.Scrollbar[0,354,550x16,val=0,vis=true,min=0,max=3469,horz]
java.awt.Scrollbar[0,23,16x331,val=0,vis=true,min=-300,max=300,vert]
java.awt.Scrollbar[534,23,16x331,val=8,vis=true,min=0,max=11,vert]
java.awt.Panel[0,0,550x23,layout=java.awt.FlowLayout]
java.awt.Label[206,5,101x13,align=left,label=Number of Samples:]
lyon.IntLabel[312,5,32x13,align=left,label=3469]
```

This list can be useful for debugging purposes. The *Frame* instance is shown in Figure 3.8.

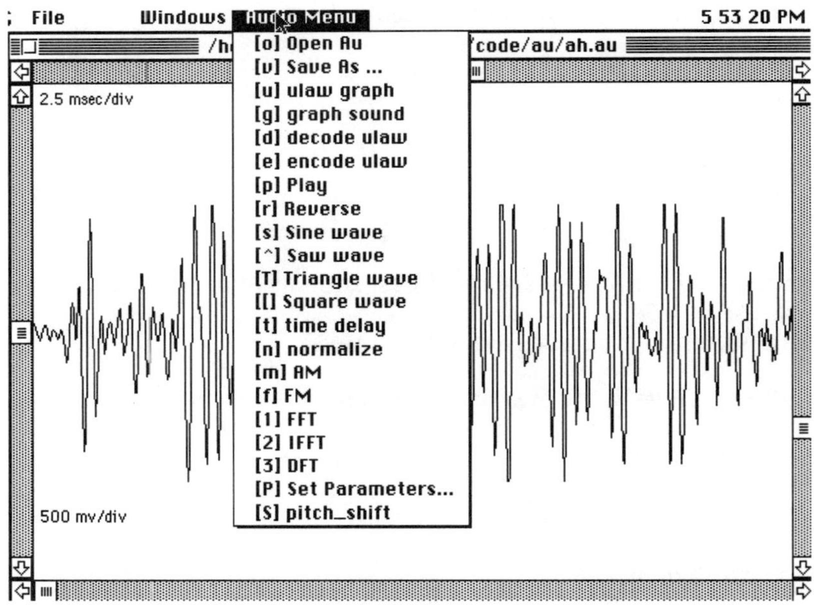

FIGURE 3.8 THE DIGITAL OSCILLOSCOPE.

The Frame Class

The *Frame* class is a descendent of the *Window* class (which in turn descends from the *Container* and *Component* classes). As an extension of the *Window* class, the *Frame* class adds a title bar, menu bar, border, cursor, and icon image. *Frame* implements *MenuContainer* and so can contain *MenuComponents*.

Class Summary

```
package java.awt;
public class Frame extends Window implements MenuContainer
    public static final int     DEFAULT_CURSOR
    public static final int     CROSSHAIR_CURSOR
    public static final int     TEXT_CURSOR
    public static final int     WAIT_CURSOR
    public static final int     SW_RESIZE_CURSOR
    public static final int     SE_RESIZE_CURSOR
    public static final int     NW_RESIZE_CURSOR
    public static final int     NE_RESIZE_CURSOR
    public static final int     N_RESIZE_CURSOR
    public static final int     S_RESIZE_CURSOR
    public static final int     W_RESIZE_CURSOR
    public static final int     E_RESIZE_CURSOR
    public static final int     HAND_CURSOR
    public static final int     MOVE_CURSOR
    public Frame()
    public Frame(String title)
    public synchronized void addNotify()
    public String getTitle()
    public void setTitle(String title)
    public Image getIconImage()
    public void setIconImage(Image image)
    public MenuBar getMenuBar()
    public synchronized void setMenuBar(MenuBar mb)
    public synchronized void remove(MenuComponent m)
    public synchronized void dispose()
    public boolean isResizable()
    public void setResizable(boolean resizable)
    public void setCursor(int cursorType)
    public int getCursorType()
}
```

Class Usage

Suppose that the following is predefined:

```
Frame f = new Frame("Title of Frame");
```

The *Frame* constructor is overloaded. You can either pass a string (which becomes the title of the frame) or you can leave the frame untitled:

```
Frame foo = new Frame();
```

```
Frame titledFrame = new Frame("Here is a title");
```

Frame instances always start life invisible and with a layout called *BorderLayout*. To get and set the title of a frame, use the following:

```
String title = f.getTitle();
f.setTitle("My title");
```

To get and set the frame's icon image:

```
Image icon = f.getIconImage();
f.setIconImage(icon);
```

To get and set the frame's menu bar:

```
MenuBar mb = f.getMenuBar();
f.setMenuBar(mb);
```

To remove the menu bar:

```
f.remove(m);
```

To dispose of the frame:

```
f.dispose;
```

To see whether the frame is resizable:

```
aBoolean = f.isResizable();
```

To set the resizablilty:

```
f.setResizable(aBoolean);
```

The cursor is set using constants that end with *_CURSOR*. To set the cursor to one of the preset cursors, use one of the following:

```
f.setCursor(DEFAULT_CURSOR);
f.setCursor(CROSSHAIR_CURSOR);
f.setCursor(TEXT_CURSOR);
f.setCursor(WAIT_CURSOR);
f.setCursor(SW_RESIZE_CURSOR);
f.setCursor(SE_RESIZE_CURSOR);
f.setCursor(NW_RESIZE_CURSOR);
f.setCursor(NE_RESIZE_CURSOR);
f.setCursor(N_RESIZE_CURSOR);
f.setCursor(S_RESIZE_CURSOR);
f.setCursor(W_RESIZE_CURSOR);
f.setCursor(E_RESIZE_CURSOR);
f.setCursor(HAND_CURSOR);
f.setCursor(MOVE_CURSOR);
```

To get the cursor:

```
anInt = f.getCursorType();
```

It is typical to handle events within a frame. Speciality frames are often used to handle specific events. As an example, we show the *ClosableFrame*.

The ClosableFrame Class

ClosableFrame is an extension of the *Frame* class that handles the close window event. It is often desirable to have windows that respond to the close event, and this event is typically handled with the following code fragment:

```
if(e.id == e.WINDOW_DESTROY) {
     hide();
     return true;
}
```

Rather than repeat the same code in every event handler of every *Frame* instance, we have devised the *ClosableFrame* class:

```
package lyon.gui;
import java.awt.*;
public class ClosableFrame extends Frame {
     // constructor needed to pass window title to class Frame
     public ClosableFrame(String name) {
          // call java.awt.Frame(String) constructor
          super(name);
     }
     // needed to allow window close
     public boolean handleEvent(Event e) {
          // Window Destroy event
          if (e.id == Event.WINDOW_DESTROY) {
               dispose();
               return true;
          }



          // it's good form to let the super class look at any unhandled events
          return super.handleEvent(e);
     } // end handleEvent()
}  // end class ClosableFrame
```

Any *Frame* that extends *ClosableFrame* will inherit the ability to handle the *WINDOW_DESTROY* event provided that the subclass returns super.*handleEvent*.

To take advantage of *ClosableFrame* we extend the it as follows:

```
...
public class AudioFrame extends ClosableFrame {
...
public AudioFrame(String name) {
        super(name);
...
```

NOTE

The constructor of the superclass is called with *super(name)* to ensure that the call to *java.awt.Frame(String)* will eventually be made.

The Panel Class

The *Panel* class is a generic extension of the *Container* class. It is often used to organize the layout of components into structured subparts, as in the example of the HTML Generator panel in Figure 3.9. The *Panel* instance has several components, including buttons, drop–down menus, and labels. These components are arranged to give the user a functional presentation of the input and output components. The *Panel* class represents a low-overhead container with its own layout manager instance. A *Panel* can also contain its own event handler, something that permits a more object–oriented approach to the event handling.

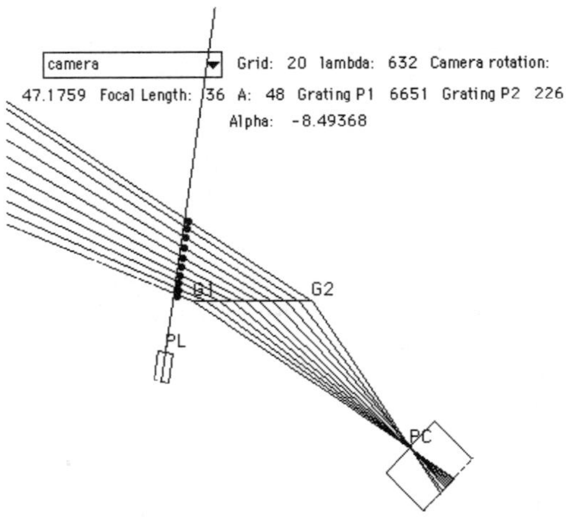

FIGURE 3.9 THE HTML GENERATOR PANEL.

Class Summary

```
 public class Panel extends Container {

public Panel()
    public synchronized void addNotify()
}
```

Class Usage

The default layout manager for the panel is *FlowLayout*. Suppose the following variable is predefined:

```
Panel p = new Panel();
```

To create a peer for the *Panel* use the following:

```
p.addNotify();
```

Beyond the *addNotify* and constructor (with default layout), the *Panel* inherits all its methods and properties from the *Container* class.

Building a Panel

In the following code we see several components (some of which will be discussed later) being added to a panel.

CD-ROM

These components are in the *TargetControlPanel.java* file, a part of the HTML Generator interface

```
package htmlconverter;
import java.awt.*;
public class TargetControlPanel extends Panel {
     Choice c;
     TargetControlPanel() {
          c = new Choice();
          c.addItem("Java");
          c.addItem("C");
          c.addItem("C++");
          c.select("Java");
          setLayout(new GridLayout(1, 3, 10, 10));
          add(new Label("Target:", Label.LEFT));
          add(c);
     }
}
```

The Checkbox Class

The *Checkbox* class extends the *Component* class. It is one of the many interface widgets that can be added to any *Container* subclass. Typically, a *Checkbox* instance is added to a *Panel* or a *Frame*. The state of the *Checkbox* instance becomes true if it is checked; otherwise, it is false. The nice thing about a *Checkbox* instance is that you do not have to handle the *Checkbox* event. *Checkbox* instances can be incorporated directly into Java programs.

Checkbox instances can be grouped into another instance called a *CheckboxGroup*, commonly called a radio button. Radio buttons are a paradigm of the push-button station–selection feature available on some car radios. Users assign buttons to their favorite and then select a station with the push of the button. The basic rule of a radio button is that you can select only one button of the group. The act of selecting one button deselects the others and one button is always selected.

Class Summary

```
public class Checkbox extends Component {
 public Checkbox()
 public Checkbox(String label)
 public Checkbox(String label, CheckboxGroup group, boolean state)
 public synchronized void addNotify()
 public String getLabel()
 public void setLabel(String label)
 public boolean getState()
 public void setState(boolean state)
 public CheckboxGroup getCheckboxGroup()
 public void setCheckboxGroup(CheckboxGroup g)
}
```

Class Usage

Suppose the following variables are predefined:

```
String label;
CheckboxGroup group;
boolean state;
Checkbox cb;
```

To make an instance of an unchecked *Checkbox* without a label or group, use this statement:

```
cb = new Checkbox();
```

To make an instance of an unchecked *Checkbox* with a label and no group:

```
cb = new Checkbox(label);
```

To make an instance of a *Checkbox* with a label, group, and known state:

```
cb = new Checkbox(label, group, state);
```

To make the peer of a *Checkbox*:

```
cb.addNotify();
```

To get and set the label of a *Checkbox*:

```
label = cb.getLabel();
cb.setLabel(label);
```

To get and set the state of the *Checkbox*:

```
state = cb.getState();
cb.setState(state);
```

To get and set the group of the *Checkbox*:

```
group = cb.getCheckboxGroup();
cb.setCheckboxGroup(group);
```

Adding Check Boxes to Frames

In *GUI.java*, the check boxes for the DiffCAD frame are placed into an array of check boxes. The array is used to manipulate the check boxes with fewer lines of code. The check boxes are shown in Figure 3.10, the DiffCAD main frame.

```
...
Checkbox ap_cb = new Checkbox("Auto-pan",null,true);
Checkbox order_cb = new Checkbox("-order",null,true);
Checkbox i3_cb = new Checkbox("-i3");
Checkbox r3_cb = new Checkbox("-r3",null,true);
Checkbox x3_cb = new Checkbox("-x3",null,false);
```

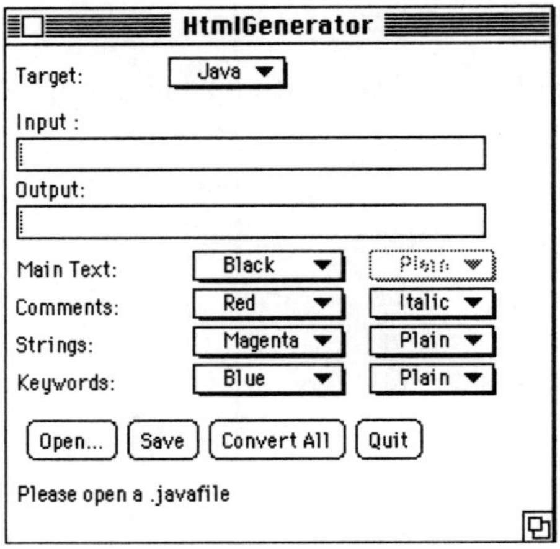

FIGURE 3.10 THE DIFFCAD MAIN FRAME. NOTE THE VECTOR PICT FILE OUTPUT.

NOTE

To use a *Checkbox* with a known state and no *Checkbox* group, you must place a *null*, as a placeholder, into the *group* parameter.

```
Checkbox checkboxes[] = {
      ap_cb,
      order_cb,
       i3_cb,
       r3_cb,
       x3_cb};
```

To add the check boxes to a frame, we use a method called *addCheckBoxes*:

```
public void addCheckBoxes( Checkbox checkboxes[]) {
     for (int i=0; i < checkboxes.length; i++) {
            add(checkboxes[i]);
            }
}
```

To perform computation with a *Checkbox*, you need only get its state. For example:

```
if (x3_cb.getState()) {
        p.x_data[i] = p.x_data[i] * -1;
       }
```

It is possible to make the *Checkbox* event trigger computation. In some cases, this mode may be the preferred one; however, if computation will lag behind the user's input rate, feedback to the user will be required and the program may seem sluggish [Tog]. Because there are several check boxes, the approach taken by DiffCAD is to update the screen with a recomputation only during a *repaint*. This policy permits us to embed *Checkbox* states directly into equations.

The Scrollbar Class

The *Scrollbar* class is a *Component* subclass. It is generally added to a *Container* class instance and is used to obtain a numeric quantity from the user. The *Scrollbar* excels at permitting a user to scroll data, such as an image or graph, in a window of limited size. The digital oscilloscope example (see Figure 3.8) has four scroll bars: top, bottom, left, and right. Their functions will be described in the "Class Usage" section. The box that slides along the scroll bar is typically called the *elevator*. Taking the elevator up is the same as clicking and holding on the box while dragging the pointing device's cursor to the top of the scroll bar.

Class Summary

```
public class Scrollbar extends Component {
public static final int     HORIZONTAL
public static final int     VERTICAL
public Scrollbar()
public Scrollbar(int orientation)
public Scrollbar(int orientation, int value, int visible, int minimum, int maxi-
mum)
public synchronized void addNotify()
public int getOrientation()
public int getValue()
public void setValue(int value)
public int getMinimum()
public int getMaximum()
```

```
public int getVisible()
public void setLineIncrement(int l)
public int getLineIncrement()
public void setPageIncrement(int l)
public int getPageIncrement()
public void setValues(int value, int visible, int minimum, int maximum)
}
```

Class Usage

Suppose that the following variables are predefined:

```
int orientation;
int visible;
int minimum,  maximum;
int value;
```

To construct a *Scrollbar* instance, you use this statement

```
Scrollbar sb;
```

The constructor defaults to a vertical scroll bar:

```
sb = new Scrollbar();
int orientation;
```

The orientation is either *Scrollbar.HORIZONTAL* or *Scrollbar.VERTICAL.*

```
sb = new Scrollbar(orientation);
int value;
int pageSize, minimum, maximum;
```

To make a *Scrollbar* instance with given orientation, value, visibility, minimum range and maximum range:

```
sb = new Scrollbar(orientation, value, visible, minimum, maximum);
```

To make a peer for the *Scrollbar* instance:

```
sb.addNotify();
```

To get the orientation of the *Scrollbar*:

```
orientation = sb.getOrientation();
```

To get and set the *value*:

```
value = sb.getValue();
sb.setValue(value);
```

Value is clipped to *Maximum* or *Minimum* if it is not between them.

To get the minimum and maximum of the *Scrollbar*:

```
value = sb.getMinimum();
value = sb.getMaximum();
```

To get the amount visible, in pixels:

```
value = sb.getVisible();
```

To set and get the elevator increment for an arrow click:

```
sb.setLineIncrement(1);
l = sb.getLineIncrement);
```

To set and get the elevator increment for a page click:

```
sb.setPageIncrement(1);
l = sb.getPageIncrement();
```

To set the values (*value* is clipped if it exceeds the valid range):

```
sb.setValues(value, visible, minimum, maximum);
 }
```

Adding Four Border Scroll Bars

In this section we show how to construct the four border scroll bars shown in Figure 3.8, the digital oscilloscope generator. An oscilloscope is a test instrument used to graph wave forms. The digital oscilloscope is intended to have some of the features of the traditional instrument (which can typically be purchased at significant cost). The following code illustrates how to add scroll bars to the *AudioFrame* (which is used to display the digital oscilloscope):

```
1. public class AudioFrame extends ClosableFrame {
2. ...
```

In lines 3–6 we create the horizontal and vertical scroll bars (bottom, top, left, and right).

```
3. Scrollbar sbHorzBottom = new Scrollbar(Scrollbar.HORIZONTAL);
4. Scrollbar sbHorzTop = new Scrollbar(Scrollbar.HORIZONTAL);
5. Scrollbar sbVertLeft = new Scrollbar(Scrollbar.VERTICAL);
6. Scrollbar sbVertRight = new Scrollbar(Scrollbar.VERTICAL);
```

For our digital oscilloscope, we want *sbVertLeft* to translate the wave form vertically. The scroll bar *sbVertRight* will scale the wave form in height. The scale factors for the wave form are given in millivolts per division. We want divisions in millivolts to reflect the steps of the physical instrument (from 5 volts to 1 millivolt). These steps become our y-scale factors. We also want the x-scale factors to alter the number of seconds used per division. To reflect the physical instrument, we proceed in steps from 0.05 microseconds per division to 5 seconds per division. This technique models the steps and range of a circa 1969 dual-trace 150–Mhz bandwidth Tektronix model 7704 portable oscilloscope. The 7704 weighs more than 50 pounds and is portable by virtue of having two handles.

```
7. private int dx = 0;
8. private int oldDx = 0;
9. private int dy = 0;
10. private int oldDy = 0;
11. private final double xScaleFactors[] = {50000, 25000, 10000, 5000, 2500, 1000,
12. 250, 100, 50, 25, 10,
13. 2.5, 1, .5, .25, .1,
14. .025, .01, .005, .0025, .001,
15. .0005};
16. private final String xSFLabels[] = {"0.05 u", "0.1 u", "0.25 u",
17. "0.5 u", "1 u", "2.5 u",
18. "5 u", "10 u", "25 u",
19. "50 u", "100 u", "250 u",
20. "500 u", "1 m", "2.5 m",
21. "5 m", "10 m", "25 m",
22. "50 m", "100 m", "250 m",
23. "500 m", "1 ", "2.5 ", "5 "};
24. private final int xsfStartIndex = 14;
25. private double xScaleFactor = xScaleFactors[xsfStartIndex];
26. private double oldXScaleFactor = xScaleFactors[xsfStartIndex];
27. private String xSFLabel = new String(xSFLabels[xsfStartIndex]);


28. private final double yScaleFactors[] = {500, 200, 100, 50, 20, 10,
29. 2, 1, .5, .2, .1};


30. private final String ySFLabels[] = {"1 m", "2.5 m", "5 m", "10 m", "25 m", "50
m",
31. "100 m", "250 m", "500 m", "1 ", "2.5 ", "5 "};
```

```
32. private final int ysfStartIndex = 8;
33. private double yScaleFactor = 1;
34. private double oldYScaleFactor = 1;
35. private String ySFLabel = new String(ySFLabels[ysfStartIndex]);
36. ...
```

```
37. public boolean handleEvent(Event e) {
38. ...
```

During the handling of the event, we can check to see whether the event target contains the same reference as one of the components.

```
39. if(e.target == sbHorzTop) {
40.       int i = sbHorzTop.getValue();
41.       xScaleFactor = xScaleFactors[i];
```

For line 42, we check to see whether the scale factor has changed. If it has not changed, we do not call *repaint*, a function that is computationally expensive.

```
42.       if (xScaleFactor != oldXScaleFactor) {
43.       xSFLabel = xSFLabels[i];
44.       repaint();
45.       oldXScaleFactor = xScaleFactor;
46.       }
47.       return true;
48. }
```

To make the scroll bars border the *AudioFrame*, we use a layout manager called *BorderLayout*:

```
public AudioFrame(String name) {
...
 setLayout(new BorderLayout());
 add("North", sbHorzTop);
 add("South", sbHorzBottom);
 add("West",  sbVertLeft);
 add("East",  sbVertRight);
```

```
 // openAudioStream will set the fileName
 // and audioStream variables.
 openAudioStream();
 // Set top scroll bar characteristics
 sbHorzTop.setValues(xsfStartIndex, 0, 0, 24);
...
```

The Label Class

Label is a subclass of the *Component* class and is generally added to a container. Labels typically are not generators of useful events. A label can be updated dynamically to reflect the state of an underlying variable. For example *GUI.java* file in DiffCAD has both static and dynamic labels. Dynamic labels are discussed in a later section.

Class Summary

```
public class Label extends Component {
    public static final int LEFT
    public static final int CENTER
    public static final int RIGHT
    public Label()
    public Label(String label)
    public Label(String label, int alignment)
    public synchronized void addNotify()
    public int getAlignment()
    public void setAlignment(int alignment)
    public String getText()
    public void setText(String label)
}
```

Class Usage

Suppose the following variables are predefined:

```
Label l;
String name;
int alignment;
```

Alignment is set to one of the following:

```
Label.LEFT, Label.RIGHT, Label.CENTER.
public class Label extends Component {
```

To construct an empty label, use this code:

```
l = new Label();
```

To construct a named label:

```
l = new Label(name);
```

To construct a named label with known alignment:

```
l = new Label(name, alignment);
```

To make the label peer

```
l.addNotify();
```

To get and set alignment:

```
alignment = l.getAlignment();
l.setAlignment(alignment);
```

To get and set the string that the label contains:

```
name = l.getText();
l.setText(name);
}
```

Adding Labels to Frames

To add labels to a frame, you need only invoke the following:

```
    add(new Label("Grid:"));
```

No event handler code is required. Also, any extension of the *Container* class can add components.

The Choice Class

The *Choice* component class provides a pop-up menu of string choices. The currently selected string becomes the title of the string menu. In many ways, a *Choice* component is like a main menu bar selection that continues to display its last selection. Pop-up menu examples can be seen in Figure 3.9, and Figure 3.10.

Class Summary

```
public class Choice extends Component {
    public Choice()
    public synchronized void addNotify()
```

```
    public int countItems()
    public String getItem(int index)
    public synchronized void addItem(String item)
    public String getSelectedItem()
    public int getSelectedIndex()
    public synchronized void select(int pos)
    public void select(String str)
}
```

Class Usage

The *Choice* component has only a single constructor. Suppose the following variables are predefined:

```
Choice choice = new Choice();
int i;
String str;
```

To create the peer, use this code:

```
choice.addNotify();
```

To find the number of strings in *choice*:

```
i = choice.countItems();
```

To get the string at location *i*:

```
str = choice.getItem(i);
```

To add a string:

```
choice.addItem(str);
```

To get the selected string:

```
str = choice.getSelectedItem();
```

To get the location of the string in the list:

```
i = choice.getSelectedIndex();
```

To select a string, by position or name, thereby making it visible:

```
choice.select(i);
choice.select(str);
}
```

Adding Choices to a Frame

In this example, we show how to add a *Choice* menu to a *Frame*.

CD-ROM

The following code is excerpted from the DiffCAD program in *GUI.java*.

The basic idea is that a graph object can be selected by choosing it from the *Choice* instance.

```
public class GUI  extends Applet implements constants  {
    Choice mouseChoice;
    String objectPropertiesStr = new String("Object Properties");
    String cameraStr = new String("camera");
    String laserStr = new String("laser");
    String laserAngleStr = new String("laser angle");
    String gratingStr = new String("grating");
    String graphPvsXStr = new String("graph p vs x");
    String graphDLvsXStr = new String("graph DL vs x");
    String allStr = new String("all");

        String mouseChoiceItems[] = {
         objectPropertiesStr,
         cameraStr, laserStr, laserAngleStr, gratingStr,
         graphPvsXStr, graphDLvsXStr, allStr
    };
```

To assist in the creation of *Choice* menus, we use the graphics utilities class Guitils:

```
public class Guitils {
     static public Choice addChoiceMenu(String items[], Container cont) {
        Choice choice  = new Choice();
        for (int i = 0; i < items.length; i++) {
             choice.addItem(items[i]);
        }
        cont.add(choice);
        return  choice;
     }   ...
```

GUI.java uses the *Guitils.addChoiceMenu* method in *create_gui*:

```
 public void create_gui() {
       setBackground(Color.white);
    Guitils.addCheckBoxes(checkboxes, this);
    mouseChoice = Guitils.addChoiceMenu(mouseChoiceItems, this);
```

Because the *Frame* class is a subclass of the *Container* class, the usage of *this* passes the *Frame*, with automatic type casting, into the *Guitils.addCheckBoxes* method. The *Guitils.addCheckBoxes* method does not keep a reference to the *Choice* instance, *choice*, because the *Container* instance will store it.

To handle the *choice*, we have adopted the policy of waiting for the mouse to be released. Then, we can process the *choice*:

```
1. public boolean handleEvent(Event e) {
2.      switch (e.id) {
3.           case Event.MOUSE_UP: {
4.                     String objectName = mouseChoice.getSelectedItem();
5.                     processMouseUp( objectName, e.x, e.y);
6.                     repaint();
7.            }
8.            case Event.MOUSE_DOWN:
9.                 anchor = new point( e.x, e.y);
```

In line 9, we store the mouse down event so that we can compute the relative mouse motion when the mouse is released.

```
10.       return super.handleEvent(e);
11.      } // switch
```

The invocation of the *processMouseUp* method causes a sequence of *if-then-else* methods to be fired:

```
1.  public void processMouseUp(String object_name, int x, int y) {
```

The *rmove* point instance in line 2 is used to compute the relative mouse motion. The upside–down nature of the coordinate system imposed by the AWT designers requires that we negate the (*y–anchor.y*) displacement. This coordinate system was probably an unfortunate design choice.

```
2.      point rmove = new point(x-anchor.x, anchor.y - y);
3.      // rmove is the relative motion of the mouse.
```

It is much faster to compare strings using a check between references. This approach is safe, because the strings were added to the *Choice* instance by reference. It is not safe to use this technique:

```
// if (object_name == "camera") // <-- this is NOT safe
```

Also, it is bad software engineering practice to embed string literals throughout the code, because typos can result in hard-to-find bugs:

```
// if (object_name.equals("camera") // <-- this is bad engineering!
```

The best way is to allow the compiler to check for type correctness. In this way, if there is a typo the compiler will emit a syntax error rather than require run-time debugging. Furthermore, the reference comparison using == is very fast.

```
        if ( object_name== cameraStr) {
             Camera c = (Camera) Geometry.camera;
             c.auto_pan = ap_cb.getState();
             c.change_config(rmove);


             Geometry.compute_rays();


        } else if ( object_name== laserStr) {
              Laser l = (Laser) Geometry.laser;
              l.change_config(rmove);
              Geometry compute_rays();
         } else if ( object_name == gratingStr) {
              Grating g = (Grating) Geometry.grating;
              g.change_config(rmove);
              Geometry compute_rays();
         }  else if ( object_name == graphDLvsXStr)
              make_graph_dl();
       else if ( object_name == graphPvsXStr)
               make_graph_p();
   else if (object_name == laserAngleStr) {
              Wedge w = (Wedge) Geometry.wedge;
              w.change_config(rmove);
              Geometry.compute_rays();
              return;
         } else if (object_name == allStr) {
         Xform.rmove(rmove);
         return;
         }
```

Summary

This chapter introduced some of the basic Java classes needed to perform basic AWT programming. In the following chapters we will present a series of subclasses that extend the *Component* class.

Chapter 4

Futil Recipes for Feudal Times

From cooking meals for hungry hired men
And washing dishes after them
—*Robert Frost*

Resistance is Futil.
—*Borg*

CD-ROM

This book comes with a package called *futils.*

The *futils* package is a collection of file utilities that simplify the writing of user-friendly GUI-based code. The *java.io* package is responsible for enabling all input-output (I/O) operations in Java. The *java.io* consists of 31 class files. Eventually, programmers will want to learn them all. However, even seasoned programmers will want to develop a set of useful tools that are less general and therefore easier to use. The premise is that simplicity is inversely related to generality and that simplicity is a feature. Thus, the *futils* package presents a set of simple tools designed to be easy to use and to make compact code that is easy to read.

File utilities are considered to be unsafe because they can rename, copy, list, and even delete files. This is all the more reason that there should be a centralized package of well-tested and well-understood classes for file manipulation. Furthermore, because of security manager restrictions, most classes in the *futils* package are not run within a browser.

The *futils* package has several public final classes, each of which has several static methods. To gain access to these methods, type the following at the head of your Java source code:

```
import futils.*;
```

The *Futil* class is a public final class that resides in the *futils* package. It contains a single private constructor that is used to prevent instantiation of the class. The *Futil* class is dependent on the *java.io* package.

The Dialog Class

An instance of a *Dialog* class is like a low-overhead *Frame* instance. A *Dialog* instance is resizable and has a title, a border, and a layout.

A *Dialog* instance has a *modal* property. When a *Dialog* instance is modal, the user is forced to interact with it. If a *Dialog* instance is not modal, the user can choose other *Window* instances (such as *Frame* instances). The default layout for a Dialog instance is *BorderLayout*.

Class Summary

```
public class Dialog extends Window {
    public Dialog(Frame parent, boolean modal)
    public Dialog(Frame parent, String title, boolean modal)
    public synchronized void addNotify()
    public boolean isModal()
    public String getTitle()
    public void setTitle(String title)
    public boolean isResizable()
    public void setResizable(boolean resizable)
}
```

Class Usage

Suppose the following variables are predefined:

```
String title;
boolean resizable, modal;
Dialog dialog;
Frame parent;
```

To make an (initially invisible) instance of a *Dialog*, use this statement:

```
dialog = new Dialog(parent, modal);
```

To make an (initially invisible) instance of a *Dialog* with a title:

```
dialog = new Dialog(parent, title, modal);
```

To create a peer:

```
dialog.addNotify();
```

To see whether a *Dialog* instance is modal:

```
modal = dialog.isModal();
```

To get and set the title:

```
title = dialog.getTitle();
dialog.setTitle(title);
```

To get and set the *resizable* property:

```
resizable = dialog.isResizable();
dialog.setResizable(resizable);
```

The FileDialog Class

The *FileDialog* class is a subclass of the *Dialog* class. Instances of *FileDialog* are used by the *futils* package to obtain file and directory information from the user. An instance of *FileDialog* will block the invoking thread when shown. The *FileDialog* class is always modal.

Class Summary

```
public class FileDialog extends Dialog
    public static final int LOAD
    public static final int SAVE
    public FileDialog(Frame parent, String title)
    public FileDialog(Frame parent, String title, int mode)
    public synchronized void addNotify()
    public int getMode()
    public String getDirectory()
    public void setDirectory(String dir)
    public String getFile()
    public void setFile(String file)
    public FilenameFilter getFilenameFilter()
    public void setFilenameFilter(FilenameFilter filter)
}
```

Class Usage

Suppose the following variables are predefined:

```
FileDialog fileDialog;
Frame parent;
String title, path;
int mode;
FileNameFilter filter;
```

Mode must be either *FileDialog.LOAD* or *FileDialog.SAVE.*

To create a *FileDialog* instance for reading a file, use:

```
fileDialog = new FileDialog(parent, title);
```

To specify a mode during instantiation:

```
fileDialog = new FileDialog(parent, title, mode);
```

To get the mode:

```
mode = fileDialog.getMode();
```

To get and set the directory:

```
path = fileDialog.getDirectory();
fileDialog.setDirectory(path);
```

To get the file name:

```
path = fileDialog.getFile();
```

To set the default file name before the dialog box is shown:

```
fileDialog.setFile(path);
```

To get and set *FilenameFilter*:

```
filter = fileDialog.getFilenameFilter();
fileDialog.setFilenameFilter(filter);
```

Futil.getReadFileName

Futil.getReadFileName is a static method that creates a file open dialog box, which is used to select a file. Once the file is selected, *Futil.getReadFileName* returns an absolute path name to the file as a *String* instance. An example of the standard file open dialog box is shown in Figure 4.1.

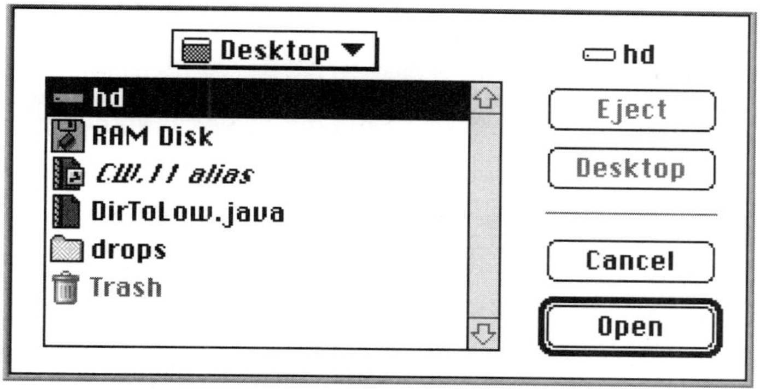

FIGURE 4.1 THE STANDARD FILE OPEN DIALOG BOX.

The appearance of the file open dialog box will be platform-dependent.

```
public static  String getReadFileName() {
  FileDialog dialog = new FileDialog(new Frame(), "select file");
  dialog.show();
  String file_name = dialog.getFile();
  String path_name = dialog.getDirectory();
  String file_string = path_name + file_name;
  System.out.println("Opening file: "+file_string);
  dialog.dispose();
  return file_string;
}
```

NOTE

The parent field of the dialog box is set to *new Frame()* so that the user does not need to pass a *Frame* instance into the *getReadFileName* parameter list. This technique simplifies the *getReadFileName* call at the expense of efficiency. The path name for the file is explicitly concatenated into the *file_string* before the return.

Typically, a programmer would prompt the user to select a file by using this statement:

```
String inputFileName = Futil.getReadFileName();
```

Futil.getWriteFileName

To get a file name to write, the mode of *FileDialog* must be set to *FileDialog.SAVE*. A sample of the save file dialog box is shown in Figure 4.2.

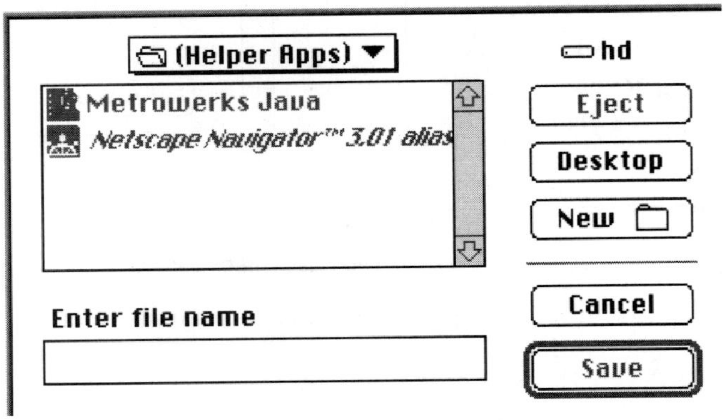

FIGURE 4.2 SAVE FILE DIALOG BOX.

This mode is set in the following code:

```
public   static  String getWriteFileName() {
       FileDialog dialog = new FileDialog(new Frame(),
      "Enter file name",FileDialog.SAVE);
      dialog.show();
      String fs = dialog.getDirectory() + dialog.getFile();
      System.out.println("Opening file: "+fs);
      dialog.dispose();
      return fs;
   }
```

To ask the user for a file string with a fully qualified path name:

```
String writeFile = Futil.getWriteFileName();
```

The File Class

The *File* class resides in the *java.io* package. An instance of the *File* class keeps track of several file-related properties, including a path to the file, information about whether the path is relative or absolute, and a series of static constants that indicate the path separator. Unfortunately, on some systems (DOS, Windows, and so on) the path name separator is represented by a backslash (\), whereas on other systems (Macintosh and UNIX variants) the path name separator is represented by a forward slash (/).

The *File* class uses a *FilenameFilter* instance, which is introduced later in the chapter.

Class Summary

```
public class File  {
   public static final String separator
   public static final char separatorChar
   public static final String pathSeparator
   public static final char pathSeparatorChar
   public File(String path)
   public File(String path, String name)
   public File(File dir, String name)
   public String getName()
   public String getPath()
   public String getAbsolutePath()
   public String getParent()
   public boolean exists()
   public boolean canWrite()
   public boolean canRead()
   public boolean isFile()
   public boolean isDirectory()
   public native boolean isAbsolute();
   public long lastModified()
   public long length()
   public boolean mkdir()
   public boolean renameTo(File dest)
   public boolean mkdirs()
   public String[] list()
   public String[] list(FilenameFilter filter)
   public boolean delete()
   public int hashCode()
   public boolean equals(Object obj)
   public String toString()
}
```

Class Usage

Suppose the following variables are predefined:

```
String absPath = Futil.getReadFileName();
// The fileName is relative and
// does not include absPath
```

```
String fileName;
String dirName; // absolute path to directory
File dirFile; // dir File instance
File destFile; // a destination file
boolean aboolean, successful;
int i;
long timeInMilliseconds; // relative time.
long bytes; // size of the file
String path; // relative or absolute
String fileNames[];
FileNameFilter filter;
```

To make an instance of the File class, use this code:

```
file = new File(absPath);
file = new File(dirName,fileName);
file = new File(dirFile,fileName);
fileName = file.getName();
```

To get the path to the file:

```
path = file.getPath();
```

To get the absolute path to the file:

```
absPath = file.getAbsolutePath();
```

To get the name of the parent directory:

```
dirName = file.getParent();
```

To see whether a file exists:

```
aboolean = file.exists();
```

To see whether a file is writable:

```
aboolean = file.canWrite();
```

To see whether a file is readable:

```
aboolean = file.canRead();
```

To see whether a *File* instance is a file or a directory:

```
aboolean = file.isFile();
aboolean = file.isDirectory();
```

To see whether *getPath* is relative:

```
aboolean = file.isAbsolute();
```

To get the relative time since modification:

```
timeInMilliseconds = file.lastModified();
```

To get the size of the file in bytes:

```
bytes = file.length();
```

To invoke *mkdir* and return true if successful:

```
successful = file.mkdir();
```

To rename to a destination file:

```
successful = file.renameTo(destFile);
```

To make all directories in this path, including the filename itself:

```
successful = file.mkdirs();
```

To list the files in a directory:

```
fileNames = file.list();
```

To use a filter to list the files in a directory:

```
fileNames = file.list(filter);
```

To delete a file:

```
success = file.delete();
```

To compute a *hashcode*

```
i = file.hashCode();
```

To see whether two files are equal:

```
aboolean = file.equals(destFile);
```

To get the *path* string:

```
path = file.toString();
```

Ls.getDirName

UNIX-type operating systems have a command called *ls*. We have modeled this command in the *futils* package using a class called *Ls*. It contains static methods and is declared as public and final. *Ls.getDirName* opens a standard file dialog box and asks the user to select a file (there is no way to select a directory as far as we know). Once the file is selected, the name of the directory that contains the file is returned:

```
static public String getDirName() {
        FileDialog fileDialog =
                 new FileDialog(new Frame(), "select file");
        fileDialog.show();
        String dirName = fileDialog.getDirectory();
        dialog.dispose();
        return dirName;
}
```

Ls.deleteFile

The *Ls.deleteFile* method takes an absolute path name (a string), converts it to a *File* instance, and then deletes the file, with error reporting to the console.

```
static public void deleteFile(String absPath) {
     File fileToDelete = new File(absPath);
     System.out.print("deleting file " + absPath);
     if (fileToDelete.exists()) {
           System.out.println(" deleted!");
           fileToDelete.delete();
           }
      else
           System.out.println(" does not exist");
}
```

Futil.getReadFile, Futil.getWriteFile, and Futil.getDirFile

The static method *Futil.getReadFile* opens a dialog box for the user and returns a *File* instance, whereas *Futil.getDirFile* works by getting a file based on the *Ls.getDirName* method:

```
public  static  File getReadFile() {
                return new File(getReadFileName());
```

```
        }
public   static  File getWriteFile() {
                return File(getWriteFileName(fs));
   }
 public   static  File getDirFile() {
                return new File(Ls.getDirName());
    }
```

An example of the File Open dialog box created by getWriteFill is shown in Figure 4.2.

The FilenameFilter Interface

FilenameFilter is an interface reference data type that requires that an *accept* method be implemented. Once the *FilenameFilter* instance is created, it can be passed to utilities that will determine whether a file name should be allowed on a list.

Class Summary

```
public interface FilenameFilter
    boolean accept(File dir, String name);
}
```

Class Usage

The best way to illustrate the usage of *FilenameFilter* is to show some examples. All the examples are taken from the *futils* package.

Typically, a more efficient approach is to keep all file names in *File* instances. There are reasons, however, for maintaining lists of files as arrays of *String* instances. This approach is particularly efficient when you're attempting to trade space for time. For example, the storage of a *String* instance is based on a dictionary that assumes the *String* is immutable, thereby permitting the reuse of substrings. Therefore, redundancy in absolute path name storage should not be very memory inefficient, because the *String* storage facility has been highly optimized. By emphasizing the storage of absolute path names in *String* instances, we have gained space efficiency. However, having to create and dispose of many *File* instances creates CPU inefficiency. Questions remain concerning the quantitative aspects of this trade-off.

DirFilter

DirFilter takes a file name and returns true if the file is a directory. It first makes an instance of the *File* class, using the file name, and then targets the instance with an *isDirectory* invocation.

```
package futils;
import java.io.*;
import java.util.*;
public class DirFilter implements FilenameFilter {
     public boolean accept(File dir, String name) {
          return new File(dir, name).isDirectory();
     }
}
```

The FileFilter Class

The *FileFilter* class, like the *DirFilter* Class, creates a temporary *File* instance to determine whether the dir+name constitutes a valid file.

```
package futils;
import java.io.*;
import java.util.*;
public class FileFilter implements FilenameFilter {
     public boolean accept(File dir, String name) {
          return new File(dir, name).isFile();
        }
}
```

The WildFilter Class

The *WildFilter* class contains the suffix string of the file name string that the programmer is looking for. Once the *WildFilter* is instantiated, the suffix is unchangeable (because of the private visibility of the suffix class variable).

```
package futils;
import java.io.*;
import java.util.*;
public class WildFilter implements FilenameFilter {
      private String suffix;
      WildFilter (String suffix_) {
           suffix = suffix_;
      }

public boolean accept(File dir, String name) {
return name.endsWith(suffix);
     }
 }
```

Ls.getWildNames

The *Ls.getWildNames* static method opens a standard file open dialog box. After a user selection, all the names are returned in a *String* array with the absolute path name.

```
static public  String[] getWildNames(String wild) {
          File dir = Futil.getDirFile();
          String absPath = dir.getAbsolutePath();
          String[] fileNames = dir.list(new WildFilter(wild));
          System.out.println("getWildNames:"+absPath);
          for (int i=0; i < fileNames.length; i++) {
          fileNames[i] = absPath+fileNames[i] ;
  }
  return fileNames;
}
```

Ls.wildToConsole

Ls.wildToConsole lists all the files to the standard output device. This is equivalent to the UNIX *ls *.wild* or the DOS *dir *.wild*.

```
static public void wildToConsole(String wild) {
  String[] files = getWildNames(wild);
  System.out.println(files.length + " file(s):");
  for (int i=0; i < files.length; i++)
       System.out.println("\t" + files[i]);
}
```

Ls.deleteWildFile

Now for the dangerous stuff: a method that deletes all the files in a directory without confirmation.

WARNING

Use this method with caution.

```
static public void deleteWildFiles(String wild) {
     String[] files = getWildNames(wild);
     System.out.println(files.length + " file(s):");
     for (int i=0; i < files.length; i++)
     deleteFile(files[i]);
}
```

Ls.WordPrintMerge

Ls.WordPrintMerge gets all the files that end with a **PICT** suffix and creates Microsoft Word print merge commands that will include the pict files in a single Word document. We used this utility in creating the figure manuscript for Chapter 1 of this book.

```
static public void WordPrintMerge() {
     String wild = "PICT";
     String[] files = getWildNames(wild);
     System.out.println(files.length + " file(s):");
     int fileNumber;
     for (int i=0; i < files.length; i++) {
          fileNumber = i + 1;
          System.out.print("Figure *."+
                     fileNumber +
                     ".        INCLUDE "hd:current:Java book:chapter I:batch 1
revl:picts:\");"
                     System.out.println(files[i]+"»");
           }
 }
```

Ls.lowerFileNames

WARNING

Ls.lowerFileNames takes a *File* instance and recursively traverses the file system, converting all file names to lowercase!

Now for the really dangerous (and fun!) stuff. There is no single command in most operating systems for doing this (for obvious reasons). *Ls.lowerFileNames* is used to maintain Web sites whose file names do not match the case of the file names in the hypertext references. This chore is particularly painful for Web masters who port Web sites from one file system to another. For example, on a DOS system the file names may appear as uppercase. On a Mac or UNIX file system, files are mixed-case.

```
public void lowerFileNames( File thePath ){
String[] fileNames = thePath.list();
String pathstr = thePath.getPath();
for( int i=0;fileNames != null && i< fileNames.length; i++ ) {
  String aFileName = fileNames[ i ];
  String newFileName = aFileName.toLowerCase();
  File theFile = new File( pathstr, aFileName );
```

```
  if( theFile.isFile() ){
  //rename theFile to lowercase
  System.out.print( i+":" + aFileName );
  theFile.renameTo( new File( pathstr, newFileName ) );
  System.out.println( "\t==>\t"+newFileName );
  }else{
  //case theFile is Dir, in the Dir,
  // repeat same procedure
  System.out.println( "Dir:"+aFileName );
  lowerFileNames( new File( pathstr+aFileName ));
  }
  }
return;
}//lowerFileNames
```

The FileOutputStream Class

The *FileOutputStream* class resides in the *java.io* package. Instances of *FileOutputStream* are used as targets of the *close* method. When an instance of a *FileOutputStream* is made, the file is opened for write access.

Because the creation and closing of a *FileOutputStream* instance can throw an *IOException*, operations involving the use of a *FileOutputStream* are surrounded by *try* and *catch*.

Class Summary

```
public class FileOutputStream extends OutputStream {
  public FileOutputStream(String name) throws IOException
  public FileOutputStream(File file) throws IOException
  public FileOutputStream(FileDescriptor fdObj)
  public native void write(int b) throws IOException;
  public void write(byte b[]) throws IOException
  public void write(byte b[], int off, int len) throws IOException
  public native void close() throws IOException;
  public final FileDescriptor getFD()  throws IOException
}
```

Class Usage

To open a file for write using a standard file open save box, define a *FileOutputStream* instance by using the *Futil.getWriteFileName()*:

```
FileOutputStream output_stream = new      FileOutputStream(getWriteFileName());
```

Suppose the following variables are predefined:

```
String fileName;
File file;
FileDescriptor fd;
byte b, bytes[];
int offset, length;
```

To open a file for write access (remember to use *try* and *catch*), use this code:

```
fos = new FileOutputStream(name);
```

To open a file using a *File* instance:

```
fos = new FileOutputStream(file);
```

To open a file using a *FileDescriptor* instance:

```
fos = new FileOutputStream(fd);
```

To write a byte of data:

```
fos.write(b);
```

To write an array of bytes:

```
fos.write(bytes);
```

To write a subarray of bytes:

```
fos.write(bytes, offset, length);
```

To close the *FileOutputStream*:

```
fos.close();
```

To get the file descriptor:

```
fd = fos.getFD();
```

Futil.getFileOutputStream

A simple way to get a file output stream—by prompting the user with a file dialog box and intercepting the possible resulting exceptions—is given next.

The *Futil.getFileOutputStream* method returns null if the output file cannot be made.

```
public static FileOutputStream getFileOutputStream() {
    FileOutputStream fos = null;
    try {fos =
            new FileOutputStream(getWriteFileName());
        }
    catch (IOException e) {
            System.out.println("futil:Could not create file");
        }
        return fos;
    }
```

Futil.closeOutputStream

As with the *getFileOutputStream*, there is a *closeOutputStream* method.

NOTE

Any subclass of the *OutputStream* (such as *FileOutputStream*) will automatically be cast into an *OutputStream* instance during the call. Thus, there is no need for a *Futil.closeFileOutputStream* method.

```
public static void closeOutputStream(OutputStream os) {
  try {os.close();} // end try
   catch (IOException exe)
   {System.out.println(
    "futil: could not close outputstream");}
  }
```

The PrintStream Class

The *PrintStream* class resides in the *java.io* package. An instance of *PrintStream* exists in *System.out* and has been introduced informally using

```
System.out.println("Hello World");
```

A *PrintStream* instance can be explicitly flushed. When flushed, the buffered output will be written. A newline can be used to trigger a flush.

Class Summary

```
public class PrintStream extends FilterOutputStream {
  public PrintStream(OutputStream out)
  public PrintStream(OutputStream out, boolean autoflush)
  public void write(int b)
  public void write(byte b[], int off, int len)
  public void flush()
  public void close()
  public boolean checkError()
  public void print(Object obj)
  synchronized public void print(String s)
  synchronized public void print(char s[])
  public void print(char c)
  public void print(int i)
  public void print(long l)
  public void print(float f)
  public void print(double d)
  public void print(boolean b)
  public void println()
  synchronized public void println(Object obj)
  synchronized public void println(String s)
  synchronized public void println(char s[])
  synchronized public void println(char c)
  synchronized public void println(int i)
  synchronized public void println(long l)
  synchronized public void println(float f)
  synchronized public void println(double d)
  synchronized public void println(boolean b)
}
```

Class Usage

Suppose the following variables are predefined:

```
OutputStream os;
PrintStream ps=new PrintStream(os);
boolean autoflush;
byte b, bytes[];
int offset, length;
boolean aboolean;
Object object;
String string;
char aChar, charArray[];
int i;
long l;
float f;
double d;
```

To make a *PrintStream* instance from any *OutputStream* instance, use:

```
ps = new PrintStream(os);
```

To automatically flush the output stream after every new line, set autoflush = true:

```
ps = new PrintStream(os, autoflush);
```

To write a byte:

```
ps.write(b);
```

To write a subarray:

```
ps.write(bytes, offset, length);
```

To flush the stream:

```
ps.flush();
```

To close the stream:

```
ps.close();
```

To flush the print stream with output stream error messages and return true if an error occurred during the life of the *PrintStream*:

```
aboolean = ps.checkError();
```

To print a newline:

```
ps.println();
```

To print an object or an object+newline:

```
ps.print(object);
ps.println(object);
```

To print a *String* or a string+newline:

```
ps.print(string);
ps.println(string);
```

To print a *char* or an array of *char*:

```
ps.print(aChar);
ps.print(charArray);
```

To print an *int*, *long*, *float*, or *double*:

```
ps.print(i);
ps.print(l);
ps.print(f);
ps.print(d);
```

To print an *int*, *long*, *float*, or *double* + newline:

```
ps.println(i);
ps.println(l);
ps.println(f);
ps.println(d);
```

To print a *boolean* or *boolean* + newline:

```
ps.print(aboolean);
ps.println(aboolean);
```

Futil.makeTocHtml

In this section we present a method for reading all the documents in a directory and creating a table of contents in HTML. The user is prompted for the input directory and location of the output file. It is beyond the scope of this book to explain HTML.

Futil.makeTocHTML uses *Futil.getDirFile* to create a standard file open dialog box that prompts the user for an HTML file.

```
1.      public static void makeTocHtml() {
2.    File dir = getDirFile();
3.    String[] files = dir.list(new FileFilter());
4.    System.out.println(files.length + " file(s):");
```

In line 5, *Futil.getFileOutputStream* is used (with exception handling, handled at a lower level).

```
5.    FileOutputStream fos = getFileOutputStream();
6.    PrintStream ps = new PrintStream(fos);
7.    ps.println("<HTML>");
8.    ps.println("<BODY>");
9.    ps.println("<ul>");
```

In lines 10 and 11, the file names are written as a part of the HTML. Note that file instances were never needed, because this is a simple text output program. The *href* is a hypertext reference that shows the file with a relative path name.

```
    for (int i=0; i < files.length; i++)
       ps.println("<LI><a href = \"" + files[i]+"\">"+
            files[i]+"</a><P>");
    ps.println("</ul>");
    ps.println("</BODY>");
    ps.println("</HTML>");
    closeOutputStream(fos);
     }
```

The browser output of *makeTocHtml* (after being run on the *futils* package) looks like this:

Cat.java
DirFilter.java
FileFilter.java
Find.java
Futil.java
Ls.java
WildFilter.java

When the user clicks on any of the underlined links, a plain-text file of the source code is shown.

The FileInputStream Class

The *FileInputStream* class resides in the *java.io* package. It can be constructed from a file name or *File* instance. *FileInputStream* is a subclass of the *InputStream* class and keeps a private copy of the *FileDescriptor* reference.

The read methods always block if there are no more bytes to read. When a read method blocks, it stops execution of the thread. This indicates that I/O bound tasks should probably be placed in their own threads. If a file will open quickly, on the other hand, placing the file I/O in its own thread may needlessly complicate the program. Thus, the decision to use threaded I/O depends on the application.

Class Summary

```
public class FileInputStream extends InputStream {
  public FileInputStream(String name) throws FileNotFoundException
  public FileInputStream(File file) throws FileNotFoundException
  public FileInputStream(FileDescriptor fdObj)
  public int read() throws IOException;
```

```
    public int read(byte b[]) throws IOException
    public int read(byte b[], int off, int len) throws IOException
    public long skip(long n) throws IOException
    public int available() throws IOException
    public void close() throws IOException
    public final FileDescriptor getFD() throws IOException
}
```

Class Usage

Suppose the following variables are predefined:

```
String fileName;
FileInputStream fis;
File file;
FileDescriptor fd;
byte b bytes[];
int length, offset;
int amountRead;
int n, numberSkipped;
```

To make an instance of the *FileInputStream* class from a file name, *File* instance, or *FileDescriptor* instance, use this code:

```
fis = new FileInputStream(fileName);
fis = new FileInputStream(file);
fis = new FileInputStream(fd);
```

To read a byte of data (-1 returned at end of stream):

```
b = fis.read();
```

To read an array of bytes (-1 returned at end of stream):

```
amountRead = fis.read(bytes);
```

To read a subarray of bytes:

```
amountRead = fis.read(bytes, offset, length);
```

To skip n bytes:

```
numberSkipped = fis.skip(n);
```

To find out how many bytes can be read (used to test the length of the file):

```
n = fis.available();
```

To close the stream:

```
fis.close();
```

To get the file descriptor:

```
fd =  fis.getFD();
```

Futil.getFileInputStream

To localize the exception handling within the *Futil* class, a static public method has been devised called *Futil.getFileInputStream*. *getFileInputStream* is overloaded to take either an absolute path name or no argument at all. When no argument is passed, a standard file dialog box is presented to the user for file selection. After a valid file is selected, *getFileInputStream* invokes the overloaded version of itself that takes a string. Should the user cancel the dialog box selection, an exception prints an error message and program execution continues.

```
public static FileInputStream getFileInputStream() {
        return  getFileInputStream(getReadFileName());
    }

public static FileInputStream getFileInputStream(String name) {
 FileInputStream fis = null;
    try
         {fis =  new FileInputStream(name);}
    catch (IOException e)
         {System.out.println("futil:Could not open file");}
         return fis;
    }
```

Recall that *FileInputStream* also has a constructor that works with a *File* instance:

```
public static FileInputStream getFileInputStream(File file) {
 FileInputStream fis = null;
    try
         {fis =  new FileInputStream(file);}
    catch (IOException e)
         {System.out.println("futil:Could not open file");}
         return fis;
    }
```

Futil.available

The *Futil.available* method is a static public method that permits a fast file check to see how many bytes are in a file. This operation was developed for the recursive *futils.DirList* class (which appears in the next section).

```
// Open the file, return -1 if file cannot be opened
// otherwise return the size in bytes
public static int available(File file) {
 FileInputStream fis = null;
 int sizeInBytes = -1;
    try {
         fis =  new FileInputStream(file);
         sizeInBytes = fis.available();
         fis.close();
    }
    catch (IOException e)
         {System.cut.println("Futil:Could not open file");}
         return sizeInBytes;
}
```

The futils.DirList Class: Recursive File Lister

The *DirList* class resides in the *futils* package. If *DirList.main* is invoked, a standard file open dialog box is presented to the user. The user selects a file, and *DirList* builds a list of all files in the directory and all subdirectories. Invocation of *DirList.main* is like the UNIX command *ls -al */* >foo* in that *DirList* opens each file, prints the size (in bytes), and stores the *File* instances, the total number of files visited, and the total number of bytes in each file. A class instance variable called *history* is of *Vector* type and holds instances of all files visited.

WARNING

The power of recursive file management is fraught with danger. Do not attempt to write recursive file deletes, as this is a serious time bomb.

```
package futils;
import java.awt.*;
import java.io.*;
import java.util.*;
import java.lang.*;
public class DirList  {
   String startDir;
```

```
public Vector history = new Vector();
int totalBytes = 0;
int totalFiles = 0;
public static void main(String args[]) {
   DirList dl = new DirList();
   dl.printStats();
}
DirList() {
   startDir = Ls.getDirName();
   startAtThisDir(startDir);
}

public void printStats() {
     System.out.println("Saw " +
             totalFiles +
             " files with a total size of " +
             totalBytes  +
             " bytes");
}
//-----------------------------
// Recursive function that given an anchor directory
// will walk directory tree
//
//-----------------------------
public void startAtThisDir(String anchorDir)   {
   FileFilter files = new FileFilter();
   DirFilter dirs = new DirFilter();
   File f1 = new File(anchorDir);
   File f2;
   FileInputStream fis;
   int i;
   String[] ls;
   int bytes;
   System.out.println("Selected -> " + anchorDir);
   System.out.println("Files in this Directory: ");
   //-----------------------------
   // Loop through all the file names in the current directory
   // and store them in history:
   //-----------------------------
   for (ls = f1.list(files), i=0;
            ls != null && i < ls.length; i++) {
        totalFiles++;
        f2 = new File(f1, ls[i]);
        bytes = Futil.available(f2);
        totalBytes +=  bytes;
        System.out.println(f2 +
             " has "+
             bytes + " bytes");
     history.addElement(f2);
   }
```

```
        //------------------------------
        // This loop recurses on all directory names in
        // the current working directory.
        //------------------------------
        for (ls = f1.list(dirs),
             i=0; ls != null && i < ls.length; i++)
           startAtThisDir(anchorDir + ls[i] + f1.separator);
    }
}
```

NOTE

The use of the *Ls.getDirName* and *Futil.available* methods lets us eliminate the try-catch structure that usually characterizes Java I/O code. As a result, the code for the *DirList* class has fewer nested code blocks. Because exceptions are handled at a lower level, some of the complexity is hidden from the programmer and the code is easier to read.

The DataInputStream Class

The *DataInputStream* class resides in the *java.io* package. An instance of the DataInputStream is a byte stream reader. *DataInputStream* instances provide-high-level methods that supply reading and casting services from a stream of bytes into various primitive data types. This class is useful when you're attempting to decode binary files (such as the audio files in Chapter 5).

DataInputStream is a subclass of *FilterInputStream* and implements the *DataInput* interface. When a read is performed and no bytes are available, the thread will block (cease to execute).

Class Summary

```
public class DataInputStream extends FilterInputStream implements DataInput {
  public DataInputStream(InputStream in)
  public final int read(byte b[]) throws IOException
  public final int read(byte b[], int off, int len) throws IOException
  public final void readFully(byte b[]) throws IOException
  public final void readFully(byte b[], int off, int len) throws IOException
  public final int skipBytes(int n) throws IOException
  public final boolean readBoolean() throws IOException
  public final byte readByte() throws IOException
  public final int readUnsignedByte() throws IOException
  public final short readShort() throws IOException
```

```
  public final int readUnsignedShort() throws IOException
  public final char readChar() throws IOException
  public final int readInt() throws IOException
  public final long readLong() throws IOException
  public final float readFloat() throws IOException
  public final double readDouble() throws IOException
  public final String readLine() throws IOException
  public final String readUTF() throws IOException
  public final static String readUTF(DataInput in) throws IOException
}
```

Class Usage

```
InputStream is;
byte b, bytes[];
int length, offset;
char c;
int i;
long l;
float f;
double d;
String string;
```

Suppose that the following variables are predefined:

```
InputStream is;
DataInputStream dis;
int numberRead;
```

To create an instance of the *DataInputStream* class, use this statement:

```
dis = new DataInputStream(is);
```

To read data into a byte array or subarray:

```
numberRead = dis.read(bytes);
numberRead = dis.read(bytes, offset, length);
```

To read *bytes.length* into *bytes* from *bytes[0]*:

```
dis.readFully(bytes);
```

To read *bytes* into a subarray:

```
dis.readFully(bytes, offset  length);
```

To skip bytes:

```
numberSkipped = dis.skip(numberToSkip);
```

To read a boolean (a single byte that is nonzero for true to be returned):

```
aboolean = dis.readBoolean();
```

To read a byte:

```
b = dis.readByte();
```

To read the byte into an *int*:

```
i = dis.readUnsignedByte();
```

To read a 16-bit signed or unsigned *short*:

```
s = dis.readShort();
i = dis.readUnsignedShort();
```

To read a 16-bit *char*:

```
c = dis.readChar();
```

To read 32 bits into an *int*:

```
i = dis.readInt();
```

To read 64 bits into a *long*:

```
l = dis.readLong();
```

To read 32 bits into a *float*:

```
f = dis.readFloat();
```

To read 64 bits into a *double*:

```
d = dis.readDouble();
```

To read a line, stopping at the end of the stream ('\n', '\r', or '\r\n'):

```
string = dis.readLine();
```

To read a Unicode Transfer Format (UTF) string:

```
string = dis.readUTF();
```

To read UTF from an *InputStream*:

```
string = DataInputStream.readUTF(is);
```

Cat.fileToStream and Cat.javasToFile

The *futils* package has a facility that works like the *cat* command of UNIX. The basic idea is that we would like Java to perform the *cat *.java >file* sequence, listing all the files in the current directory into a single file.

On line 2 we build a list of files with the **.java** suffix. Lines 6 and 7 pass the file name and print stream of each file to the *Cat.fileToStream* method.

```
1.static public void javasToFile() {
2.String[] files = Ls.getWildNames("java");
3.   FileOutputStream fos =
4.   Futil.getFileOutputStream();
5.   PrintStream ps = new PrintStream(fos);
6.   for (int i=0; i < files.length; i++)
7.   fileToStream(files[i], ps);
8.   Futil.closeOutputStream(fos);
9. }
10.static public void fileToStream(String fileName, PrintStream output) {
11.System.out.println("cat: "+fileName);
12.FileInputStream fis = Futil.getFileInputStream(fileName);
13.    String line;
```

The mapping is such that there may be many input streams but only a single output stream. This requires that we open and close the *FileInputStream* many times on several different files.

NOTE

The *try* and *catch* surrounding lines 15–17 reflect the *IOException* that line 16 might throw.

```
14.    try {
15.        DataInputStream dis = new DataInputStream(fis);
16.      while ((line = dis.readLine()) != null)
17.      output.println(line);
18.     } // try
19.     catch (Exception exe)
20.      {System.out.println("cat:Error on input file");}
21.   Futil.closeInputStream(fis);
22.  }
```

The DataOutputStream Class

Like the *DataInputStream* class, the *DataOutputStream* class resides in the *java.io* package. *DataOutputStream* is a byte stream writer, providing high-level methods that supply writing services from various primitive data types into a stream of bytes. This class is useful when you're attempting to encode binary files (such as the audio files in Chapter 5).

DataOutputStream is a subclass of *FilterOutputStream* and implements the *DataOutput* interface.

A *DataOutputStream* instance keeps track of the number of bytes that it has written. This value is kept in protected storage. All output performed with blocking is blocked until the data is finally written. If the output does not say "with blocking," the output can be buffered and flushed either manually or automatically.

A UTF-8 format is used to write strings in a machine-independent manner. *Chars* from 1 to 127 are written as a single byte. *Chars* from 128 to 2,047 and 0 are written by a pair of bytes. *Chars* from 2,048 to 65,535 are written by three bytes. The actual bit format is given in [Gosling and Yellin].

Class Summary

```
public class DataOutputStream extends FilterOutputStream implements DataOutput.
  public DataOutputStream(OutputStream out)
  public synchronized void write(int b) throws IOException
  public synchronized void write(byte b[], int off, int len)
  public void flush() throws IOException
  public final void writeBoolean(boolean v) throws IOException
  public final void writeByte(int v) throws IOException
  public final void writeShort(int v) throws IOException
  public final void writeChar(int v) throws IOException
  public final void writeInt(int v) throws IOException
  public final void writeLong(long v) throws IOException
  public final void writeFloat(float v) throws IOException
  public final void writeDouble(double v) throws IOException
  public final void writeBytes(String s) throws IOException
  public final void writeChars(String s) throws IOException
  public final void writeUTF(String str) throws IOException
  public final int size()
}
```

Class Usage

Suppose that the following variables are predefined:

```
char c;
String string;
OutputStream os;
short s;
int i, offset, length;
long l;
float f;
double d;
byte b;
byte bytes[];
```

To construct a new *DataOutputStream* instance, use this code:

```
DataOutputStream dos = new DataOutputStream(is);
```

To write a byte, with blocking:

```
dos.write(b);
```

To write a subarray of bytes:

```
dos.write(bytes, offset, length);
```

To flush the output stream:

```
dos.flush();
```

To write a boolean (as a 0 or 1 byte):

```
dos.writeBoolean(aboolean);
```

To write a byte to an underlying 8-bit representation:

```
dos.writeByte(b);
```

To write a *short* to an underlying 16 bit representation with the high byte first:

```
dos.writeShort(s);
```

To write a *char* to an underlying 16-bit representation with the high byte first:

```
dos.writeChar(c);
```

To write an *int* to an underlying 32-bit representation with the high byte first:

```
dos.writeInt(i);
```

To write a *long* to an underlying 64-bit representation with the high byte first:

```
dos.writeLong(l);
```

To write a *float* to an underlying 32-bit representation with the high byte first:

```
dos.writeFloat(f);
```

To write a *double* to an underlying 64-bit representation with the high byte first:

```
dos.writeDouble(d);
```

To write a string as a sequence of bytes (this casts each *char* as a byte and then writes the byte; it will write *s.len* bytes):

```
dos.writeBytes(s);
```

To write a string as a sequence of 16-bit *chars*. This will output 2 * *s.len* bytes:

```
dos.writeChars(s);
```

To write a string in a machine-independent UTF-8 format:

```
dos.writeUTF(s);
```

To get the number of bytes written:

```
i = dos.size();
```

Java, C, C++ -> HTML: The HTML Converter

Formats! You want formats? We got formats! Formats for browsers (HTML). Formats for word processors (RTF). Formats for computer languages (Java, C, C++). Formats for images (JPEG, GIF, PPM). Formats—you can't live with 'em, you can't live without 'em!

In this section we address the conversion of Java, C, and C++ into formatted and colored HTML.

CD-ROM

One of the packages included with DiffCAD is the *htmlconverter* package.

The *htmlconverter* package contains the support systems for the *HtmlGenerator* class, a Java, C, and C++ source code conversion program that generates formatted and colored HTML files. The following code was converted to HTML by *HtmlGenerator*. The HTML was then converted into RTF (rich text format), the format Microsoft Word reads. The conversion was performed using a program called MacLinkPlus [DataViz].

```
package htmlconverter;
import java.io.*;
```

JavaStream extends *DataInputStream* and so inherits the *DataInputStream* methods.

```
public class JavaStream extends DataInputStream
implements JavaText, CText, CplusplusText {
JavaHtmlString mainText = new JavaHtmlString();
JavaHtmlString comments = new JavaHtmlString();
JavaHtmlString strings = new JavaHtmlString();
JavaHtmlString keywords = new JavaHtmlString();
DataOutputStream dos;
public static String[] reservedWords = javaReservedWords;
```

The *JavaStream* constructor takes an *InputStream*.

```
JavaStream(DataInputStream s0, DataOutputStream s1) {
    super(s0);
    dos = s1;
}
public void convertToHtml() {
    int index;
    boolean isCommentedOut = false;
    boolean isQuoted = false;
    boolean isSingleQuoted = false;
```

The basic idea behind the *convertToHtml* method is that a string of characters should be read using the *readln* method. Because *JavaStream* is a subclass of *DataInputStream*, *JavaStream* inherits the *readln* method.

```
String line;
     String str1, str2;

     try {
          while (true) {
```

Line 5 reads until the end of a line and places the characters into the *line* string.

```
5.                  line = readLine();      // get a line
6.                  if (line == null) {          // if line is null
7.                          break;                 // break this loop
8.                  }
```

After a line of text is read, a series of search-and-replace operations is performed. For example:

```
1.      for (int counter = 0; counter < line.length(); counter++) {
```

Recall from Chapter 2 that to find *aString* starting from *anInt* we use the following:

```
anInt = anotherString.indexOf(aString, anInt);
```

So in line 2, we start from the *counter* position and look for the less-than character.

```
2.      index = line.indexOf("<", counter);
3.      if (index < 0) break;
4.      // if < is not found then break this loop
```

We then insert the string at the index point, replacing the '<' with '&lt'. This replacement is performed because HTML represents the '<' character as '&lt' and holds '<' as reserved.

```
5.      line = line.substring(0, index) + "&lt;"
6.              + line.substring(index + 1, line.length());
7.      // replace < to &lt;
8.      counter = index + 4;
9.      // update counter
10.      }
```

This search-and-replace is continued until all the replacements are performed. Then the line is written to the *DataOutputStream* instance.

StreamTokenizer

The *StreamTokenizer* class resides in the *java.io* package. It is used to convert an *InputStream* instance into a stream of tokens. Refer to [Gosling and Yellin] for a more comprehensive description of *StreamTokenizer*. Bytes are read from the *InputStream* instance and are treated as unsigned *shorts* that range from 0 to 255.

A *StreamTokenizer* instance is an instance of a parser. The parser settings can be altered to recognize C-style comments, C++-style comments, and line terminators and to convert tokens to lowercase.

StreamTokenizer is a higher-level *Stream* than *DataInputStream*.

Class Summary

```
public class StreamTokenizer {
  public int ttype
  public static final int TT_EOF
  public static final int TT_EOL
  public static final int TT_NUMBER
  public static final int TT_WORD
  public String sval
  public double nval
  public StreamTokenizer (InputStream I)
  public void resetSyntax()
  public void wordChars(int low, int hi)
  public void whitespaceChars(int low, int hi)
  public void ordinaryChars(int low, int hi)
  public void ordinaryChar(int ch)
  public void commentChar(int ch)
  public void quoteChar(int ch)
  public void parseNumbers()
  public void eolIsSignificant(boolean flag)
  public void slashStarComments(boolean flag)
  public void slashSlashComments(boolean flag)
  public void lowerCaseMode(boolean fl)
  public int nextToken() throws IOException
  public void pushBack()
  public int lineno()
  public String toString()
}
```

Class Usage

Suppose the following variables are predefined:

```
String sval;
double nval;
InputStream is;
int low, hi, ch, token, lineNumber;
String s;
StreamTokenizer st;
```

To make an instance of a *StreamTokenizer*, use this code:

```
st = new StreamTokenizer(is);
```

StreamTokenizer has several constants held as public static final *ints*. They are used by case statements to determine the token type. They are the end-of-file and end-of-line tokens:

```
StreamTokenizer.TT_EOF
StreamTokenizer.TT_EOL
```

To determine whether the token is a number (value in *nval*) check the Return of St. next Token() to see if it is equal to:

```
StreamTokenizer.TT_NUMBER
```

To determine whether the token is a word (value in *sval*) check St. next Token() for a return of:

```
StreamTokenizer.TT_WORD
```

To make all characters ordinary:

```
st.resetSyntax();
```

To cumulatively specify the Unicode characters to be used for words:

```
st.wordChars(low, hi);
```

To spec the whitespace *chars*:

```
st.whitespaceChars(low,hi);
```

Ordinary *chars* are returned by *nextToken*. To spec the ordinary *chars*:

```
st.ordinaryChars(low, hi);
```

To add an *ordinaryChar*:

```
st.ordinaryChar(ch);
```

To add a single-line comment *char* (all *chars* to the end of line are comment *chars*, and the tokenizer skips them):

```
st.commentChar(ch);
```

To spec the *quoteChar* to delimit a string:

```
st.quoteChar(ch);
```

To spec that numbers should be parsed:

```
st.parseNumbers();
```

To have *TT_EOL* returned by *nexttoken*:

```
flag = true;
st.eolIsSignificant(flag);
```

To select '/*' comments:

```
st.slashStarComments(flag);
```

To select '//' comments:

```
st.slashSlashComments(flag);
```

To select automatic lower case conversion:

```
lowerCaseMode(true);
```

To parse a token, returning *ttype*:

```
ttype = st.nextToken();
```

To push a token back into the stream:

```
st.pushBack();
```

To get the current line number:

```
i = st.lineno();
```

To convert the *StreamToken* to a string:

```
s = st.toString();
```

For an example of use, please see the next section.

Futil.readDataFile

To save the state of our DiffCAD program, we write several key parameters into a file. These parameters are stored with keywords and values so that a user with a text editor (perhaps a nonprogrammer) can alter the data. A sample data file follows:

```
Data format is order dependentlyon.Laser     Rotation= 34.4198          6.47311
-36.7812 lyon.Camera     rho=      18.4931     pc=              125.3     -102.35
A=      4.8      F=      3.6 lyon.Wedge      p1 =      -403.685     591.244 lyon.Grating
P1=      6651      P2=      2261      L=      81
```

The *readDataFile* method is in the *futils* package. It is used to take a file name and read data into an array of *doubles*. To perform this task we use a *StreamTokenizer* instance called *tokens*.

```
1. public static  void readDataFile(String file,double data[]) {
2.     System.out.println("processing:\t" + file);
3.     FileInputStream inputFile =
4.                                getFileInputStream(file);
5.     StreamTokenizer tokens = new StreamTokenizer(inputFile);
6.     int next = 0;
7.     int num = 0;
```

Note that the *tokens.nextToken()* must be nested in a try-catch block.

```
8.     try {
9.          while ((next = tokens.nextToken()) != tokens.TT_EOF) {
10.                switch (next) {
11.                     case tokens.TT_WORD:
12.                          break;
```

Lines 11 and 12 indicate that words are ignored (they are for humans!). Only numbers matter, and because people are warned not to change the order of the parameters, future versions of the data file can add numbers only to the end of the file. They cannot insert numbers.

```
13.                     case tokens.TT_NUMBER:
14.                          data[num] = (double) tokens.nval;
15.                          System.out.println(num+": "+ data[num]);
16.                          num = num + 1;
17.                          break;
18.                     case tokens.TT_EOL:
19.                          break;
20.                }
21.          }
22.     }
23.     catch (Exception exe)
24.          {System.out.println("listFilteredHrefFile:er!");}
25.     closeInputStream(inputFile);
26. }
```

Creation of the file is a simple matter. In the following code, we open a file as a *PrintStream* instance, write the data, and then close the file:

```
public static void save()  {
        FileOutputStream os = Futil.getWriteFileOutputStream();
        PrintStream output = new PrintStream(os);
        output.println("Data format is order-dependent");
        Geometry.print(output);
        Futil.closeOutputStream(os);
  }
```

The *Geometry.print* is a public static method that uses the *Print* class in the *futils* package, which is described in the following section.

Futil.Print

To ease the implementation of output to files and the console, the futils package provides its own *Print* class. This class is a public final class that has a private constructor to prevent instantiation. The key point is that the *Print* class stores an instance of *PrintOutputStream* in a settable variable. Thus, output can be directed to a file or, by default, to the console. The trick is that the change of the *PrintOutputStream* is centralized, and all methods that use the *Print* class benefit from this centralization. For example, in the *Geometry* class we write the following:

```
1. static public void print(PrintStream out) {
```

We invoke the *setPrintStream* method in the *Print* class, invoke *print*, and then reset the stream.

```
2. Print.setPrintStream( out);
3.          print();
4. Print.setPrintStream( System.out);
5.    } // print
```

Every shape knows how to print (using the *Print* class) and so will use the new output stream when printing. If the *print* method of line 6 is invoked without line 2 being executed, the output will be directed to the console.

```
6. static public void print() {
7.                laser.print();
8.                camera.print();
9.                wedge.print();
10.               grating.print();
11.               System.out.println("Lambda = "+ lambda.getValue());
12.               System.out.println("s = " + s());
13.    } // print
```

The code for *futils.Print* follows:

```
    package futils;
    import java.util.*;
    import java.io.*;
    public final class Print {
      // prevent instantiation
private Print() {}
private static     PrintStream output = System.out;
public static void setPrintStream(PrintStream ps) {
        output = ps;
    }
public static PrintStream getPrintStream() {
        return output;
    }
public static void d(double i)
        {Print.output.print(i+"\t");}
public static void ln(double d)
        {Print.output.println(d);}
public static void ln(String str)
        {Print.output.println(str);}
public static void print(String str)
        {Print.output.print(str);}
public static void print(double d)
        {Print.output.print(d);}
public static void print(int d)
        {Print.output.print(d);}
public static void className(Object o)
        {Print.output.print(o.getClass().getName() + "\t");}
}
```

Futil.writeFilteredHrefFile

A buggy Web authoring tool (such as Netscape 3.01 Gold) will produce HTML that has hrefs (hypertext references) that contain embedded spaces. The API documentation shows an href on line 8. The space between "API" and "Documentation" is a bug. Browsers will stop reading the href name after the first space. To get browsers to recognize the space, we must replace spaces with their hexadecimal equivalent, '%20'.

```
1. <html>
2.     <head>
3.     <title>
4. API User's Guide
5.     </title>
6.     </head>
7.     <body>
8.     <a href= "API Documentation/packages.html" > Java API</a>
```

The corrected version of line 8 follows:

```
<a href= "API%20Documentation/packages.html" > Java API</a>
```

To perform this transformation, we build a custom tokenizer that looks for quoted strings and replaces all spaces that they contain with '%20'. This approach has the bug of also transforming the spaces in non-href strings to '%20' (a bug we have been able to live with).

```
1. public static void writeFilteredHrefFile(String inputName, String outputName) {
2. System.out.println("Filtering:\t" + inputName +"\t>\t"+outputName);
3. try {
```

Lines 4–7 make an input stream and a *StreamTokenizer* instance.

```
4.      FileInputStream is = new FileInputStream(inputName);
5.      StreamTokenizer tokens = new StreamTokenizer(is);
6.      FileOutputStream os = new FileOutputStream(outputName);
7.      PrintStream output = new PrintStream(os);
8.      int i;
9.      int next = 0;
```

The interesting bit is line 10. Recall that *resetSyntax* will make all characters ordinary. This means that there are no comment characters.

```
10.     tokens.resetSyntax();
```

Line 11 specifies that all characters are word characters. Line 12 identifies the only character of interest, the quote character. Recall that *nextToken* will read until the *quoteCharacter*, setting *sval* to the value of the body of the string contained in quotes.

```
11.     tokens.wordChars(0,255);
12.     tokens.quoteChar('"');
13.     while ((next = tokens.nextToken()) != tokens.TT_EOF) {
14.        switch (next) {
15.        case '"':
16.             output.print('"');
```

We now output a quote and then scan the string for a space. If a space is found, we output a "%20"; otherwise, we output the character.

```
17.         for (i=0;i<tokens.sval.length();i++)
18.                if (tokens.sval.charAt(i) == ' ')
19.                     output.print("%20");
```

```
                 else
                      output.print(tokens.sval.charAt(i));
                    output.print('"');
                    break;
```

In all other cases we output the string read. This technique is nice, because we read only up to the point at which a delimiting character occurs.

```
         case tokens.TT_WORD:
                    output.print(tokens.sval+" ");
                    break;
         case tokens.TT_NUMBER:
                    output.print(tokens.nval+" ");
                    break;
         case tokens.TT_EOL:
                    output.println();
                    break;
              } // end switch
         } // end while
      is.close();
      os.close();
      } // end try
      catch (Exception exe)
       {System.out.println("writeFilteredHrefFile:er!");}
}
```

Exercises

The astute reader will note that we did not perform local handling of exceptions for *StreamTokenizer*. We suggest the development of a *StreamTokenizer* with local handling of exceptions as an exercise.

Another topic for exploration is to extend the *futils.DirList* class. One useful feature might be to include a constructor that takes a *String* instance for a suffix. *DirList* could then construct a *WildFilter* instance to form a list of only those files whose ending matches the suffix string.

A topic we have not yet seen in any book is the creation of a UNIX-like command-line interface to Java that would let us type commands such as *ls *.java > foo* into the command-line reader. This could be an excellent way to employ piping and to write a series of data-processing–type tools.

Summary

In this chapter we covered the *Dialog* and *FileDialog* classes as the widgets to use to get file names from the user. *Futils.getReadFileName* and *Futils.getWriteFileName* have embedded calls to the *FileDialog* constructor, so there is no reason to embed a file name into a file in the Java source code. The *futils* package contains a list of classes that are used daily and save several lines of source code each time they appear in a program.

From a design viewpoint, the programmer might object to handling exceptions locally, as the *futils* classes attempt to do. This approach generally does not appear to be a problem. In short, we have found the local handling of exceptions to be the correct approach in all cases that we have encountered so far.

This chapter introduced a number of new classes that are unique to this book. The *Ls* class has the ability to list batches of files using the '*ls *.<suffix>*" and place the results into internal data structures. The *Ls* class provides the ability to recursively traverse the directory tree structure using a form that is like the UNIX '*ls-al*/**'. The *Cat* class takes the output of *Ls* and lists the files into a single file. Batch processing was also used by the *HtmlGenerator* class to input C, C++ or Java and output colored and formatted HTML, with a hyper-linked HTML index.

Chapter 5

Digital Audio Processing

> Thy voice sounds like a prophet's word;
> And in its hollow tones are heard
> —*Fitz-Greene Halleck. 1790–1867.*

What Is Digital Signal Processing?

Sound is a pressure wave that traverses a medium. Sound pressure waves in air are the objective cause of human hearing. Sound will not travel through a vacuum, but it will travel through various phases of matter (solid, liquid, and gas).

A *transducer* is a device that takes power from one system and supplies power to another system. For example, a microphone is a transducer that takes sound power and supplies electrical power. The electrical power supplied by the microphone forms an analog signal. An analog signal is continuous.

Digitization is a process that converts continuous signals into a digital form. Digitization (also known as analog-to-digital conversion) is performed by sampling and quantization. *Sampling* is the process of converting a continuous signal into a set of voltages. *Quantization* is the process of converting the sampled voltages into a countable set of digital values. Analog data that is converted to digital data is said to be PCM-encoded. PCM stands for pulse code modulation and is a broad term that can refer to any type of digital encoding of analog data. Figure 5.1 depicts a PCM encoder.

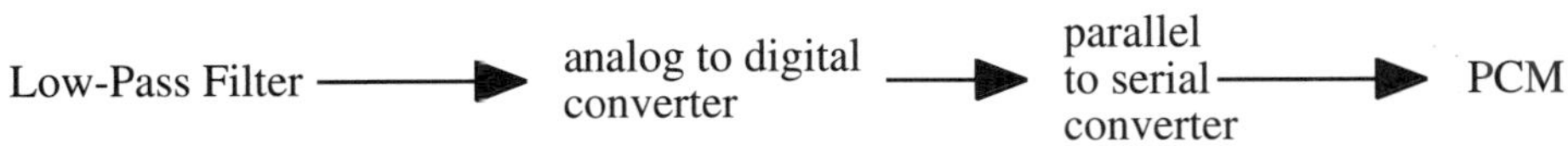

FIGURE 5.1 BLOCK DIAGRAM OF A PCM ENCODER.

A *low-pass* filter (called an anti-aliasing filter) is typically set to attenuate frequencies at or above one-half the analog-to-digital converter's sampling rate (this is known as the Nyquest frequency).

To transform the PCM signal back into the analog domain, we couple a digital-to-analog converter with another low-pass filter. A block diagram of the PCM decoder is shown in Figure 5.2.

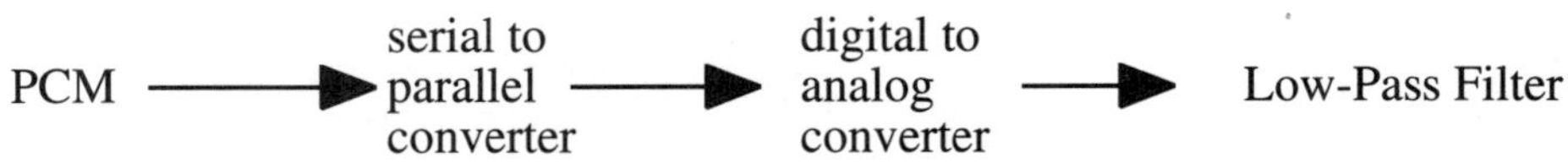

FIGURE 5.2 BLOCK DIAGRAM OF A PCM DECODER.

Digital signal processing is a kind of data processing that operates on PCM data. Thus, broadly speaking, audio, image, and image sequence processing are one-dimensional, two-dimensional, and three-dimensional digital signal processing.

In common usage, *digital signal processing* refers to one-dimensional signals, ***V***(*t*). In image processing we often speak about two-dimensional signals, ***I***(*x*,*y*). This chapter deals only with one-dimensional digital signal processing in Java.

Why Do We Need Digital Signal Processing?

A digital signal stream may come from any energy (sound, measurement, temperature, speed, pressure, radiation, and so on). Non-physical phenomena can also produce a digital stream of data (financial data, statistical data, network traffic, and so on).

In short, digital signal processing can be performed on any recordable event. Digital signal processing is a kind of data processing.
In this chapter we treat only the restricted domain of audio digital signal processing in Java. There are several reasons for this:

- Java can already play audio files.
- The techniques can be extended to other types of data.
- We can hear the results.
- It is fun!

What Is the Spectrum of a Signal?

The harmonic content of a signal is called its *spectrum*. The spectrum of a signal consists of a series of sin and cosine waves. A French mathematician, Jean Baptiste Joseph de Fourier (1768–1830), showed that harmonic waves (sine and cosine waves) can be summed in a series to form any periodic waveform. The summation (called the *superposition principle*) fails to approximate a waveform when the equations governing the waveform are nonlinear (shock waves, turbulence, chaos, and so on) [Halliday]. The series was first formulated by, and is used in, harmonic analysis (also called Fourier analysis). Harmonic analysis is the process that determines the harmonic components of a complex wave. The series can be written as

$$v(t) = a_0 + (a_1 \cos t + b_1 \sin t) + (a_2 \cos 2t + b_2 \sin 2t) + \ldots \qquad (5.1)$$

where $a0$, $a1$, $b1$, $a1$, $b2$, are constants called Fourier coefficients.

For example, a sawtooth wave can be computed by letting k in the following go to infinity:

$$f(x) = \frac{2}{\pi}\sum_{n=1}^{K}\frac{(-1)^{(n-1)}\sin(n\pi x)}{n}$$

When $k = 10$, the waveform of Figure 5.3 results.

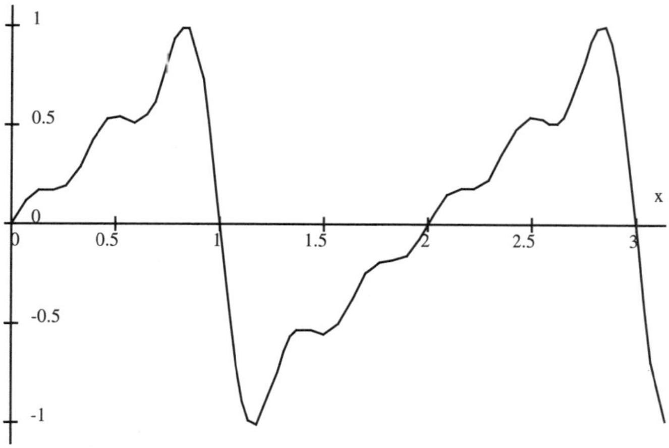

FIGURE 5.3 SAWTOOTH WAVEFORM WITH K = 10

When $k = 100$, the waveform of Figure 5.4 is produced.

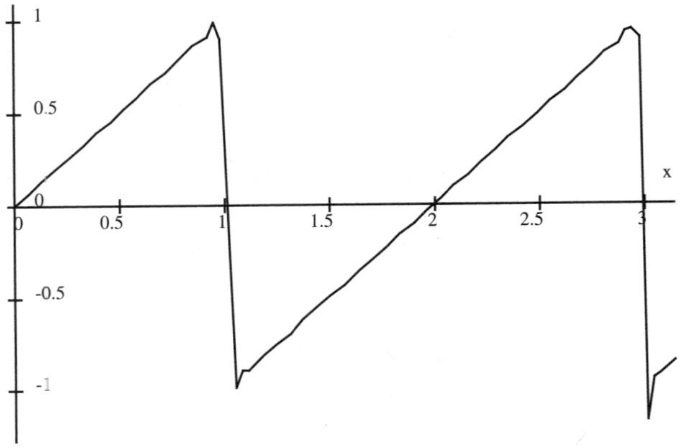

FIGURE 5.4 SAWTOOTH WAVEFORM WITH K=100.

When the waveform to be approximated is not periodic, the summation is replaced by the Fourier transform:

$$V(f) = F[v(t)] = \int_{-\infty}^{\infty} v(t)e^{-2\pi ift}dt \tag{5.2}$$

$$v(t) = F^{-1}\left[V(f)\right] = \int_{-\infty}^{\infty} V(f)e^{2\pi ift}dt \tag{5.3}$$

Where *ei(* is given by Euler's identity:

$$e^{i\theta} = \cos\theta + i\sin\theta \tag{5.4}$$

Euler's identity can lead to several equivalent representations for the Fourier series. For example, there is the sine-cosine representation:

$$x(t) = \sum_{n=0}^{\infty} a_n \cos(2\pi n f_0 t) + \sum_{n=1}^{\infty} b_n \sin(2\pi n f_0 t)$$

where f_0=frequency and f_0 = nth harmonic of f_0. The constants, known as Fourier coefficients, are found by correlating the time dependent function, *x(t)*, with an Nth harmonic sine-cosine pair:

$$a_0 = \frac{1}{T}\int_0^T x(t)dt$$

$$a_n = \frac{2}{T}\int_0^T x(t)\cos(2\pi nf_0 t)dt$$

$$b_n = \frac{2}{T}\int_0^T x(t)\sin(2\pi nf_0 t)dt$$

Another common representation of the Fourier series is the amplitude-phase representation. This is also a result of Euler's identity:

$$\mathrm{x(t)} = \mathrm{c}_0 + \sum_{n=1}^{\infty} c_n \cos(2\pi f_0 t + \theta_n)$$

$$c_0 = \frac{1}{T}\int_0^T x(t)dt$$

$$c_n = \sqrt{a_n^2 + b_n^2}$$

$$\theta_n = -\tan^{-1}\left(\frac{b_n}{a_n}\right)$$

In general, the use of the representation of the Fourier transform is a matter of preference, because the various representations are equivalent. The Fourier transform has some interesting properties, and although it is beyond our scope to state them all (or prove any of them), we give some of them here.

A Fourier transform representation of a periodic signal has discrete spectral components at *f*0 and *nf*0, as shown by Equation 5.1. An a periodic signal has a continuous and infinite spectrum, as shown by Equations 5.2 and 5.3. This means that time-limited signals (which are, by definition, aperiodic) have infinite bandwidth.

The effective bandwidth of a signal is the width of the spectrum that contains the most power. The average power in a given interval of time is computed by

$$P = \frac{1}{t_1 - t_2} \int_{t_1}^{t_2} |x(t)|^2$$

To compute the average power in a periodic signal whose period is *T*, we use

$$P = \frac{1}{T} \int_0^T |x(t)|^2 \, dt$$

The PSD (power spectral density) is the power at a specific frequency, *S* (*f*). Chapter 6 will discuss the computation of the PSD in more detail.

The Fourier transforms are important because they permit computation in either the time domain or the frequency domain. Some relations with Fourier transforms follow.

Superposition states that linear combinations in the time domain become linear combinations in the frequency domain:

$$a_1 V_1(f) + a_2 V_2(f) = F[a_1 v_1(t) + a_2 v_1(t)] \tag{5.5}$$

Delay in the time domain causes a phase shift in the frequency domain:

$$V(f) e^{-2\pi i f} = F(v(t - t_d)) \tag{5.6}$$

Scale change in the time domain causes a reciprocal scale change in the frequency domain:

$$\frac{1}{|\alpha|} V\left(\frac{f}{\alpha}\right) = F(v(\alpha t)), \alpha \neq 0 \tag{5.7}$$

The convolution theorem states that multiplication in the time domain causes convolution in the frequency domain:

$$V * W(f) = F(v(t) w(t)) \tag{5.8}$$

Where convolution between two functions of the same variable is defined by

$$V * W(f) \equiv \int_{-\infty}^{\infty} V(\lambda)W(f-\lambda)d\lambda$$

(5.9)

For proofs of these results, see [Carlson].

What Does Sampling Do to the Spectrum of a Signal?

The digitization process, as depicted in Figure 5.1, starts with the process of sampling. The sampling-reconstruction process need not perform digitization and yet is responsible for the introduction of additional harmonic content into the reconstructed signal. The sampling-reconstruction process is shown in Figure 5.5.

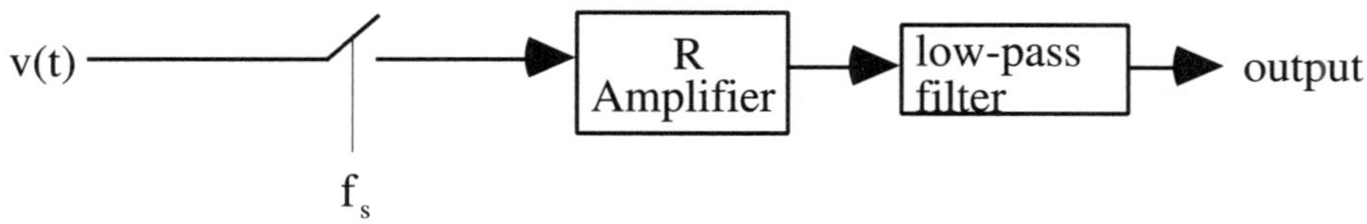

FIGURE 5.5 THE SAMPLING-RECONSTRUCTION PROCESS.

Figure 5.5 shows a continuous signal, *v(t)* , being sampled at a rate of *fs* using an electronically controlled switch. The output of the switch is fed into an amplifier that has a gain of *R*. Mathematically, the sampling function can be expressed in terms of a Dirac delta function multiplied by *v(t)* in the time domain. The Dirac delta function, *((t)*, is a generalized function that is defined by

$$\int_{-\varepsilon}^{\varepsilon} \delta(t)dt = 1$$

(5.10)

Where (is arbitrarily small. Thus, the Dirac delta function is a unit impulse that occurs at time *t*=0. We can make the Dirac delta function fire at any time, t_d, by using the form

$$\delta(t - t_d) \tag{5.11}$$

Furthermore, we can model the sampling function as an infinite sum of Dirac delta functions:

$$s(t) = \sum_{n=-\infty}^{\infty} \delta(t - n / f_s) \tag{5.12}$$

The switching function multiplies *s(t)* by *v(t)*. The sampled function is represented by

$$v_s(t) = v(t)s(t) = v(t) \sum_{n=-\infty}^{\infty} \delta(t - n / f_s) \tag{5.13}$$

Recall that multiplication in the time domain is convolution in the frequency domain:

$$V_s(f) = F[v_s(t)] = V(f) * S(f) \tag{5.14}$$

The Fourier transform of an impulse train is also an impulse train.

$$V_s(f) = V(F) * \sum_{n=-\infty}^{\infty} f_s \delta(f - nf_s) \tag{5.15}$$

Finally, we see that sampling a signal at a rate of *fs* causes the spectrum to be reproduced at f_S intervals.

$$V_s(f) = f_s \sum_{n=-\infty}^{\infty} V(f - nf_s) \tag{5.16}$$

A sample of this spectrum replication is shown in Figure 5.6.

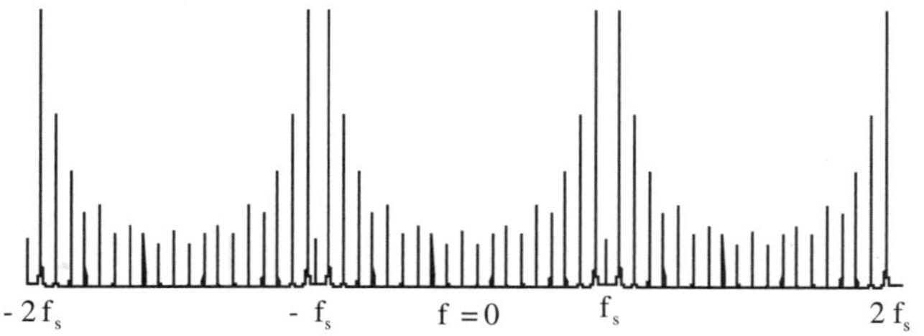

FIGURE 5.6 SPECTRUM REPLICATION DUE TO SAMPLING.

Equation 5.16 is called the aliasing formula.

Audio Files

In this section we describe how to read, play, graph, and write audio files. The current Java API can open and play an audio file but not decode it. The code we present in this chapter extends and builds on the existing Java API.

CD-ROM

You will find a collection of sound utilities for the Mac at http://wwwhost.ots.utexas.edu/mac/pub-mac-sound.html. The book's CD-ROM has a collection of sound utilities that are useful for recording and resampling audio files.

The sun.audio Package

AudioData, *AudioDataStream*, *AudioPlayer*, *AudioStream*, *AudioStreamSequence*, and *ContinuousAudioDataStream* are public classes that reside in the *sun.audio* package.

WARNING

Code that uses classes in the Java *sun.audio* package does so in an undocumented and unsupported way. When the package changes, code that depends on its classes may not work.

NOTE

A trick of the Java trade is to unzip a library of files (such as the **classes.zip** file that comes with some applet viewers. Unzipping creates a large number of class files (446 for JDK 1.0.2), and it is a good idea to perform this operation on a copy of the file (don't break your applet viewer!). Using a class browser (such as the *class dumper* included with DiffCAD), you can list the fields of the class.

A clip of audio data contains µ–law-encoded, 8-bit, 8,000-Hz sample rate data. This data can be used to construct an *AudioDataStream* instance. The *getAudioData* method gets an audio clip from the cache. *AudioStream* is an input stream used to play *AudioData*. For more documentation on the *sun.audio* package, see http://www.cdt.luth.se/java/doc/sun/shared/Package-sun.audio.html.

NOTE

Current implementations have been tested with the book software on the MacOS System 7.5x, Windows 95/NT, and Sun Solaris operating systems. The sound player software cannot make audio output on a Wintel platform that lacks a sound card. Furthermore, some sound cards do not get driven properly by the current Java sound drivers. Soundblaster-compatible cards seem to work with this book's software.

We can only guess how classes in the *sun.audio* package might work. There is no documentation on which to base a solid code foundation. However, the code that we present in the following section works for now.

The AudioStream Class

The *AudioStream* class resides in the *sun.audio* package. It is used to convert an *InputStream* to an *AudioStream*. The *AudioStream* class extends the *FilterInputStream* class. As a subclass of *FilterInputStream*, the *AudioStream* class inherits all the methods and attributes of *FilterInputStream*.

Class Summary

We cannot reverse-engineer the software that makes up *AudioStream*, so we reproduce the raw class dumper output from the DiffCAD program:

```
package sun.audio
import java.lang.Thread;
import java.io.IOException;
/*
 * This class has 1 optional class attributes.
 * These attributes are:
 * Attribute 1 is of type SourceFile
 *     SourceFile : AudioStream.java
 */


 public class AudioStream extends FilterInputStream {
  public void AudioStream(java.io.InputStream a);
  public int read(byte a[], int b, int c);
  public sun.audio.AudioData getData();
  public int getLength();
}
```

Class Usage

Suppose the following variables are predefined:

```
InputStream is;
int position, length;
byte bytes[];
AudioData ad;
```

To construct an instance of the *AudioStream* class (throws *IOException*), use this code:

```
AudioStream as = new AudioStream(is);
```

To construct an instance of an *AudioStream* from a URL:

```
AudioStream as = new AudioStream(url.openStream());
```

To read an array of bytes from the input stream starting at *position* and proceeding for *length* bytes (throws *IOException*):

```
as.read(bytes, position, length);
```

To find the length of the *AudioStream* instance:

```
length = as.getLength();
```

To get the *AudioData* from an *AudioStream* instance:

```
ad = as.getData();
```

The AudioData Class

The *AudioData* class resides in the *sun.audio* package. An instance of *AudioData* is often called a *clip* of audio data. The *AudioData* instance consists of 8-bit µ-law-encoded, 8,000-Hz sample rate data. An instance of the *AudioData* stream is used to construct an instance of an *AudioDataStream*. An instance of an *AudioDataStream* can be played to the speaker of audio-capable machines.

WARNING

Because *AudioData* is a resident of the *sun.audio* package, the interface is subject to change without notice. The class summary was derived from the DiffCAD class dumper.

Class Summary

```
public class AudioData extends java.lang.Object {
public void AudioData(byte a[]);
}
```

Class Usage

Suppose the following variables are predefined:

```
AudioStream as;
byte bytes[];
```

To make an instance of the *AudioData*, use this code:

```
ad = new AudioData(bytes);
```

To get data from an *AudioStream*:

```
ad = as.getData();
```

The AudioDataStream Class

The *AudioDataStream* resides in the *sun.audio* package. It extends *ByteArrayInputStream*. An *AudioDataStream* instance provides an input stream to play *AudioData*.

WARNING

Because *AudioDataStream* is a resident of the *sun.audio* package, the interface is subject to change without notice. The class summary was derived from the DiffCAD class dumper.

Class Summary

```
public class AudioDataStream extends ByteArrayInputStream {
  public void AudioDataStream(sun.audio.AudioData a);
}
```

Class Usage

AudioDataStream has only one public method, its constructor.

Suppose the following variables are predefined:

```
byte ulawData[];
AudioData audioData = new AudioData ( ulawData);
```

To make an instance of *AudioDataStream*, use this code:

```
audioDataStream = new AudioDataStream (audioData);
```

Reading and Playing an AU File

The following code fragment, from the *UlawCodec* class on the accompanying CD-ROM, shows how to read data from a Sun AU file. This code is used to make and play an *AudioDataStream* instance:

```
package lyon.audio;
import java.io.*;
import sun.audio.*;
import futils.Futil;
public class UlawCodec implements Runnable {
private byte ulawData[];
private double doubleArray[];
private  AudioDataStream audioDataStream = null;
private  AudioStream audioStream = null;
....
// ------------------------------
//
// public ReadAUFile(String name);
//
// This constructor can be used to acquire the audio samples
// from an AU file
//
// ------------------------------
 public void readAUFile(String fileName){
      // force recomputation
      // of the doubleArray, because it is invalid
      doubleArray = null;
  try {
   audioStream = new AudioStream(new FileInputStream(fileName));
   // AudioStream constructor expects data stream from AU file as input
   ulawData = new byte[audioStream.getLength()];
   audioStream.read(ulawData, 0, audioStream.getLength());
  } catch(Exception e){}
 }
 public void readAUFile() {
      String fileName = Futil.getReadFileName();
      readAUFile(fileName);
 }
```

The AudioStreamSequence Class

The *AudioStreamSequence* class resides in the *sun.audio* package. An instance of *AudioStreamSequence* can be used to convert a sequence of input streams into a single *InputStream*. This class is typically used to play two audio clips in sequence.

WARNING

Because *AudioStreamSequence* is a resident of the *sun.audio* package, the interface is subject to change without notice. The class summary was derived from the DiffCAD class dumper.

Class Summary

```
public class AudioStreamSequence extends java.io.InputStream {
  public void AudioStreamSequence(java.util.Enumeration a);
  public int read();
  public int read(byte a[], int b, int c);
}
```

Class Usage

Suppose the following variables are predefined:

```
AudioStream as1, as2;
Vector v = new Vector();
v.addElement(as1);
v.addElement(as2);
int lenghtRead, length, position;
byte bytes[];
```

To make a new instance of the *AudioStreamSequence*, use this code:

```
AudioStreamSequence ass = new AudioStreamSequence(v.elements());
```

To play the *AudioStreamSequence* instance:

```
AudioPlayer.player.start(audiostream);
```

To read from an instance of *AudioStreamSequence* and flip to the next stream if an EOF is encountered (throws *IOException*):

```
lengthRead = ass.read(bytes, position, length);
lengthRead = ass.read();
```

The AudioPlayer Class

The *AudioPlayer* class resides in the *sun.audio* package. This class provides an interface to play an instance of *AudioStream*.

WARNING

Because *AudioPlayer* is a resident of the *sun.audio* package, the interface is subject to change without notice. The class summary was derived from the DiffCAD class dumper.

Class Summary

```
import java.io.PrintStream;
import java.lang.System;
public class AudioPlayer extends java.lang.Thread {
public static final sun.audio.AudioPlayer player;
private void AudioPlayer();
public synchronized void start(java.io.InputStream a);
public synchronized void stop(java.io.InputStream a);
public void run();
}
```

Class Usage

Suppose the following variables are predefined:

```
AudioStream as;
AudoData ad = as.getData();
```

To play an audio stream, use this code:

```
AudioPlayer.player.start(as);
```

To stop playing an audio stream:

```
AudioPlayer.player.stop(as);
```

The μ–law CODEC Concept

In this section we show how to use a software CODEC (coder-decoder) to decode and play a Sun AU audio file. The Java API uses a μ-law (pronounced "mu-law") compression technique for playing audio files. The supported μ-law format consists of logarithmically companded, 8-kHhz sample rate, byte-quantized, voice-grade audio. The word *compandor* is a contraction of compressor and expander [BTL]. The sample time for an 8,000 samples per second system is 1/8,000 = 0.000125 second = 125 μs. Such a format generates an 8,000 sample/second * 8 bits / sample = 64 kbps data stream and is common in telephony. Also, the peak bandwidth of such a compression format is as follows:

$$Bandwidth = \frac{Sampling\ Rate}{2} = 4khz \qquad (5.17)$$

The International Telecommunication Union (ITU), formerly CCITT, has created a specification called G.711. Two PCM algorithms are defined in the G.711 standard: A-Law and μ-law. In both the A-Law and μ-law format, the sample rate is 8 kHz. In a linear PCM system, there are uniform voltage quantization steps [Bates].

A copy of the G.711 specification is available for sale at http://www.itu.int/itudoc/itu-t/rec/g/g700-799/g711_27434.html (it cost 20 Swiss francs the last time we checked). The Sun reference implementation is on the accompanying CD-ROM along with a wealth of other audio information. The μ-law encoding formula is given by

$$y = F(x) = sign(x)V_{max}\frac{\ln\left(1+\frac{\mu x}{V_{max}}\right)}{\ln(1+\mu)}$$

where

$$-V_{max} \le x \le V_{max} \qquad (5.18)$$

The A-Law formula is given by

$$y = \frac{A}{1+\ln A}\left(\frac{x}{V_{max}}\right), \left|\frac{x}{V_{max}}\right| < \frac{1}{A}$$

and

$$y = \frac{sign(x)}{1+\ln A}\left(1+\ln\left(\frac{Ax}{V_{max}}\right)\right), \frac{1}{A} \le \left|\frac{x}{V_{max}}\right| \le 1 \quad (5.19)$$

Typical values of the compression parameters used in Equations 5.18 and 5.19 are as follows:

$$\mu = 100 \text{ and } 255$$
$$A = 87.6 \quad (5.20)$$

A graph of Equation 5.18, with *V*max =1 and μ = 255, is shown in Figure 5.7. The value μ=255 is used for North American telephone transmission. The value *A*=87.6 is used for European telephone transmission. The rationale for companding is that the human ear cannot differentiate between amplitudes of sound waves as the amplitude increases [Embree].

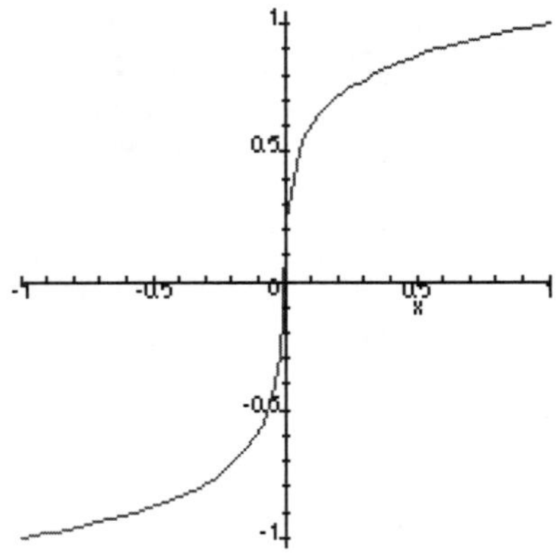

FIGURE 5.7 GRAPH OF THE M-LAW TRANSFER FUNCTION WITH M=255.

One metric of PCM performance is the ratio of the signal power to quantization noise power (signal-to-quantizing distortion ratio). The basic idea behind A-

Law and μ-law companding is that a logarithmic curve can be used to improve the signal-to-quantizing distortion ratio at low signal levels.

Binary PCM will have a number of quantization levels given by the following:

$$N_q = 2^{N_b} \tag{5.21}$$

where

Nq = the number of quantization levels

Nb = the number of bits per sample

The signal power varies from 0 to 1, inclusive, and is given by

$$S \in [0 \ldots 1]$$

The signal voltage, x, varies from

$$-V_{\max} \le x \le V_{\max} \tag{5.22}$$

For simplicity we assume that

$$V_{\max} = 1 \tag{5.23}$$

The uniform quantizer divides the signal voltage evenly among the number of quantization levels. The uniform quantizer's quantization error voltage is given by

$$-\frac{1}{2^{N_b}} \le \varepsilon \le \frac{1}{2^{N_b}}$$

where

ε = quantization error voltage (5.24)

The average quantization error is 0, but the root-mean-square (RMS) value of the quantization error is the mean square quantization noise power. The term *root-mean-square* refers to the fact that the computation is performed by taking the square root of the mean of error voltage squared [Carlson and Gisser]. Recall that for a continuous probability distribution function, the expectation is taken by

$$E[X] = \int_{-\infty}^{\infty} x f_X(x)dx$$

and the variance is taken by

$$\sigma_X^2(t) = E\left[|X(t)|^2\right]$$

The variance expands to

$$\sigma_X^2(t) = \int_{-\infty}^{\infty} x^2 f_X(x)dx$$

Also recall that for a discrete random variable,

$$E[X] = mean = \frac{1}{N}\sum_{i=1}^{N} x_i$$

and

$$\sigma_X^2(t) = \text{var}iance = E\left[|X(t)|^2\right] = \frac{1}{N}\sum_{i=1}^{N}\left(x_i - E[X]\right)^2$$

Thus, we compute the RMS error by integrating the square of the quantization error voltage over the range in Equation 5.24, assuming a zero mean:

$$\sigma^2 = \frac{1}{\left(2/N_q\right)}\int_{-1/N_q}^{1/N_q} \varepsilon^2 \, d\varepsilon = \frac{1}{3N_q^{\ 2}} \tag{5.25}$$

When the maximum signal voltage is constrained, as in Equations 5.22 and 5.23, we compute the signal-to-quantization noise power as

$$SNR_D = 10\log_{10}\left(3\times 2^{2N_b} S_x\right)$$
$$SNR_D = 10\log_{10} 3 + 20N_b \log_{10} 2S_x \quad (5.26)$$

Where the signal power is given by S_X. If the signal power is equal to 1, the range on the upper bound for the signal-to-quantization noise power is

$$SNR_D \le 4.8 + 6N_b \quad (5.27)$$

With the 8-bit PCM system and uniform quantization, the best we can hope for is a SNR of 52.8 dB. Note that the SNR (in dB) falls off linearly as a function of the power in Equation 5.26. It can be shown that the companding equations of Equations 5.18 and 5.19 will provide an improvement in the SNR when the signal power falls below –20 dB.

For signal powers greater than –20 dB, companding degrades performance relative to uniform quantization, assuming that the PDF (probability distribution function) of a voice signal has a Laplace distribution of the form

$$p(x) = \frac{1}{2} \propto e^{(-\propto x)}$$

[Carlson].

The compression parameter values given in Equation 5.20 are based on an assumed PDF of an input signal. The PDF assumption is required for telephony applications. However, in the instance of audio files that are stored on static media (such as a CD-ROM or Web server hard drive), the PDF can be computed off-line. For such a system, the SNR is

$$SNR_D = \frac{S_x}{\sigma^2} = \frac{3N_q^{\,2} S_x}{K_z}$$

where

$$K_z = 2\int_0^1 \frac{p(x)}{\left[y'\right]^2} \quad (5.28)$$

computing the derivative of Equation 5.18 with respect to x and substituting into Equation 5.28 yields

$$K_z = 2 \int_0^1 \frac{p(x)\,(V_{max} + \mu\, x)^4 \ln(1+\mu)^4}{\mu^4\, V_{max}^{\;4}}\, dx \tag{5.29}$$

Once the PDF, *p(x)*, is computed, the companding parameter can be precomputed using a criterion of optimality based on the bit rate budget. For example, a bit rate budget of 16 kbps might require a 4-bit sample with a 4-kHz sampling rate. Such a system might be used to stream audio (transmit in real time) via a low-data-rate phone connection. Ideally, a server should be able to resample and compand the audio before transmission in response to the bit rate budget.

The UlawCodec Class

The *UlawCodec* class is a public class that resides in the *lyon.audio* package.

CD-ROM

The *UlawCodec* class performs the μ-law CODEC function in addition to providing file save and open services.

Java can, in principle, process any file format. The basic problem, however, is that Java's current API can play only an *AudioDataStream* that is μ-law - encoded. Even though Java can read, process, and write any audio data format file, an intermediary μ-law encoding is required to play the audio.

Class Summary

```
public class UlawCodec implements Runnable {
  public UlawCodec()
  public UlawCodec(String name)
  public UlawCodec(short linearArrayOfShort[])
  public UlawCodec(double linearArrayOfDouble[])
  public UlawCodec(byte ulawArrayOfByte[])
  public void readAUFile(String fileName)
```

```
  public void readAUFile()
  public void writeAUFile(String fileName)
  public void writeAUFile()
  public void playSync()
  public void playAsync()
  public byte [] getUlawData()
  public void setUlawData(byte ulawArrayOfByte[])
  public double[] getDoubleArray()
  public int getLength()
  public double getDuration()
  public void reverseUlaw()
  public static void main(String argc[])
}
```

Class Usage

The *UlawCodec* class has several constructors. Each must, as its main goal, construct a μ-law-encoded byte array in a private storage area. The only way to obtain access to this storage area is by the *getUlawData* and *setUlawData* methods, in part because of a series of parallel data structures that must maintain their consistency. For example, when you invoke the *getDoubleArray* method, a check is performed to see whether the internal *DoubleArray* is null. If the array is null, it is set using computations involving *ulawData*. The consistency maintenance mechanism is invisible to the programmer.
Suppose the following variables are predefined:

```
UlawCodec ulc;
String fileName;
byte ulawArrayOfByte[];
short linearArrayOfShort[];
Double linearArrayOfDouble[];
int length;
double timeInSeconds;
String args[];
```

To read a Sun AU file using a standard file open dialog box, use this code:

```
ulc = new UlawCodec();
```

To read in a Sun AU file using a file name:

```
ulc = new UlawCodec(fileName);
```

To construct a *UlawCodec* instance from a 16-bit linear data array:

```
ulc = new UlawCodec(linearArrayOfShort);
```

To construct a *UlawCodec* instance from a linear double array:

```
ulc = new UlawCodec(linearArrayOfDouble);
```

To overwrite the internal data and read a new AU file, given a file name:

```
ulc.readAUFile(fileName);
```

To prompt the user for a file name and overwrite the internal data:

```
ulc.readAUFile();
```

To write the internal data as a new AU file, given a file name:

```
ulc.writeAUFile(fileName);
```

To prompt the user for a file name and then write a Sun AU file:

```
ulc.writeAUFile();
```

To play synchronously, returning only after the sound is played:

```
ulc.playSync();
```

To play asynchronously, returning right away and playing the sound in the background:

```
ulc.playAsync();
```

To get the raw companded byte data:

```
ulawArrayOfByte = ulc.getUlawData();
```

To set the raw companded byte data:

```
ulc.setUlawData(ulawArrayOfByte);
```

To get the data as an array of linear doubles:

```
linearArrayOfDouble = ulc.getDoubleArray();
```

To get the number of samples:

```
length = ulc.getLength();
```

To get the play time in seconds:

```
timeInSeconds = ulc.getDuration();
```

To reverse the *Ulaw* data, forcing recomputation of the *DoubleArray*:

```
ulc.reverseUlaw();
```

To test the read, play, and write methods:

```
UlawCODEC.main(args);
```

Reading and Writing μ-law

The following example, excerpted from the *UlawCodec.java file*, shows how the main method is implemented:

```
1. public static void main(String argc[]){
2.   UlawCodec ulc = new UlawCodec();
3.   ulc.playSync();
4.   ulc.writeAUFile();
5. }
```

Line 2 shows the default constructor for the CODEC. The default constructor opens the standard file dialog box to let the user select an AU file. Line 3 plays the sound and does not return until the sound has completed playing. The *writeAUFile* opens a dialog box, where the user must type a file name to save the AU file.

The Oscillator Class

The *Oscillator* class is a public class in the *lyon.audio* package and is independent of all other packages. Several instances of the *Oscillator* class can be made to create banks of oscillators. The *Oscillator* class uses double-precision data arrays and can make very low distortion waveforms. The weak link in the chain is the Java API's voice-grade audio realization. To combat this limitation (before

Sun does) would require the development of Java native methods for playing audio on all platforms. Such a development effort requires many hours of skilled labor.

Class Summary

```
public class Oscillator {
 public Oscillator(double frequency, int length)
 public double[] getSineWave()
 public double[] getSquareWave()
 public double[] getSawWave()
 public double[] getTriangleWave()
 public double getDuration()
 public int getSampleRate()
 public double getFrequency()
 public void setModulationIndex(double I)
 public void setModulationFrequency(double fm)
 public double[] getFM()
 public double[] getAM()
}
```

Class Usage

The *Oscillator* class has a number of private properties that are accessed via get and set methods. An *Oscillator* instance is created for a fixed carrier frequency and number of samples. All waveforms vary from –1 to 1. Suppose the following variables are predefined:

```
double frequency = 440;
double length = 2000; // the total number of samples
Oscillator osc;
double audioData[];
double timeInSeconds;
int sampleRate;
double indexOfModulation;
```

To make an instance of an *Oscillator*, use this code:

```
osc = new Oscillator(frequency, length);
```

To get sine, square, sawtooth, and triangle waves:

```
audioData = osc.getSineWave();
audioData = osc.getSquareWave();
audioData = osc.getSawWave();
audioData = osc.getTriangleWave();
```

To get the time the wave form will last, in seconds:

```
timeInSeconds = osc.getDuration();
```

To get the sample rate in Hz:

```
sampleRate = osc.getSampleRate()
```

To get the frequency in Hz:

```
frequency = osc.getFrequency()
```

To set the index of modulation of the FM oscillator:

```
osc.setModulationIndex(indexOfModulation);
```

To set the modulation frequency of both the AM and FM oscillators:

```
osc.setModulationFrequency(frequency);
audioData = osc.getFM()
audioData = osc.getAM()
```

The index of modulation and the modulation frequency are covered in more detail in Chapter 6.

Class Examples

The *AudioFrame* class generates a series of waveforms using the *Oscillator* class and the *UlawCodec* class.

```
public class AudioFrame extends ClosableFrame {
     private     UlawCodec ulc;
     private     Oscillator osc =
          new Oscillator(440,4000);
...
```

```
public class AudioFrame extends ClosableFrame {
     private     UlawCodec ulc;
     private     Oscillator osc =
          new Oscillator(440,4000);
...
public void play() {
          ulc.playSync();
}
public void sineWave() {
     ulc = new UlawCodec(
          osc.getSineWave());
     play();
}
public void squareWave() {
     ulc = new UlawCodec(
          osc.getSquareWave());
     play();
}
public void sawWave() {
     ulc = new UlawCodec(
          osc.getSawWave());
     play();
}
public void triangleWave() {
     ulc = new UlawCodec(
          osc.getTriangleWave());
     play();
}
public void am() {
     osc.setModulationIndex(0.5d);
     osc.setModulationFrequency(200d);
     ulc = new UlawCodec(
          osc.getAM());
     play();
}


public void fm() {
     osc.setModulationIndex(0.5d);
     osc.setModulationFrequency(200d);
     ulc = new UlawCodec(
          osc.getFM());
     play();
}
```

Class Implementation

An *Oscillator* generates repeated waveforms by constructing a single cycle of the waveform into a wavetable. The wavetable is copied repeatedly into an array of double data known internally as *audioData.*

The *Oscillator* class is implemented with a series of private class variables:

```
1.     package lyon.audio;
2.     import futils.utils.Computation;
3.     public class Oscillator {
4.     private double audioData[];
5.     private double waveTable[];
```

Lines 4 and 5 show that *audioData waveTable* are unallocated until the constructor is invoked. Line 6 shows the *sampleRate.* The constructor could be overloaded to take other sample rates, but 8,000 Hz is the default for Java's current API.

```
6.     private int sampleRate = 8000;
```

Line 7 shows the *frequency*, in Hz. For a sine wave, this is the number of wavetable cycles per second that must be clocked out. *Lambda* is the number of seconds in one cycle. Line 9 shows the number of samples in a single cycle of the wavetable, if this were computed with precision. Keep in mind that the length of a wavetable is always an integer, and as a result the *samplesPerCycle* must be converted.

```
7.     private double frequency;
8.     private double lambda;
9.     private double samplesPerCycle;
```

Oscillator construction requires that the carrier frequency and number of cycles be known. Line 3 shows the memory allocation for *audioData.*

```
1.     public Oscillator(double frequency_, int length) {
2.     frequency = frequency_;
3.     audioData = new double[length];
```

Once the period of the waveform is computed, on line 5, we can compute the number of samples in a cycle of the wavetable. This is the *samplesPerCycle* variable, which is cast into an integer. Given the integral approximation, the computation of the actual frequency at which the wavetable is clocked out is as follows:

$$f_{actual} = sampleRate / waveTable.length \tag{5.30}$$

Where the wavetable length is:

$$waveTable.length = round(sampleRate / f) \tag{5.31}$$

To compute the error in the digital oscillator instance, we subtract the frequency that we wanted from the rate cycles that are clocked out of the wavetable.

$$f_e = f - f_{actual}$$

For example, for a frequency of 440 Hz, *waveTable.length* = 18, *sampleRate* = 8,000 *audioData.length* = 4,000, and the actual frequency = 444.444 Hz. Exact frequencies can result when the desired frequency is an exact multiple of the *sampleRate.* For example, 400 will be reproduced with precision because 8,000/400 = *a waveTable.length* of 20.

```
    //the period of the wave form is
    lambda = 1/frequency;
    //The number of samples per period is
    samplesPerCycle = sampleRate * lambda;

     delta_freq = 1/samplesPerCycle;
     waveTable =
          new double[(int) samplesPerCycle];


     }
```

Building the WaveTable

The *AudioDataFromTable* method in the *Oscillator* class is used to turn a single cycle of *WaveTable* into a long array of audio data. A constraint on the *audioData* array is that it must have an absolute value that is strictly less than 1 (because of the companding formulas).

```
1. private double[] AudioDataFromTable() {
2.      int k = 0;
3.      for (int i = 0; i < audioData.length; i++) {
```

Line 4 builds the *audioData* from the *waveTable*. The indexes, *i* and *k* , both begin at 0. Lines 6 and 7 reset *k*, while index *i* continues to increment.

```
4.           audioData[i] = waveTable[k];
5.           k++;
6.           if (k >= waveTable.length)
7.                k = 0;
8.      }
9.      System.out.println("\nlambda="+lambda+
10.          "\nfrequency = "+frequency+
11.          "\nwaveTable.length = "+waveTable.length+
12.          "\nsampleRate = "+sampleRate+
13.          "\naudioData.length = "+audioData.length+
14.          "\nactual frequency = "+actualFrequency());
15.      return audioData;
16.      }
```

In the *getSineWave* method, the *waveTable* is computed for a single cycle. To make sure that the absolute value of the amplitude of the sine wave is always less than 1, it is first multiplied by 0.98 (line 20).

```
17.      public double[] getSineWave() {
18.      for (int i=0; i<waveTable.length; i++)
19.           waveTable[i] =
20.                0.98*Math.sin(twopi * i/waveTable.length);
21.      return AudioDataFromTable();
22.      }
```

To build a wavetable for a saw wave, we set the initial voltage to –1 and then compute a change in voltage, *dv* , using the length of the wavetable, *L* so that

$$V_0 = -1, dv = \frac{2}{L}$$

Then, after L–1 intervals, the final voltage will reach a value of 1–dv. The following code implements the *getSawWave* method:

```
1.      public double[] getSawWave() {
```

In line 2, the initial voltage is set to a value that is a little higher than 1.0 because of the constraint on the CODEC's input. Also, in line 4, the check is i<*waveTable.length* rather than i <= *waveTable.length*. Thus, the saw wave will end at (*v*–d*v*) rather than at 1.0 volts.

```
2.      double v = -0.99;
3.      double dv = 2.0 / (double) waveTable.length;
4.      for (int i=0; i<waveTable.length; i++){
5.           waveTable[i] = v;
6.           v += dv;
7.      }
8.      System.out.println("Sawwave ends at:"+(double)(v-dv));
9.      return  AudioDataFromTable();
10.      }
```

The saw wave output is shown in Figure 5.8.

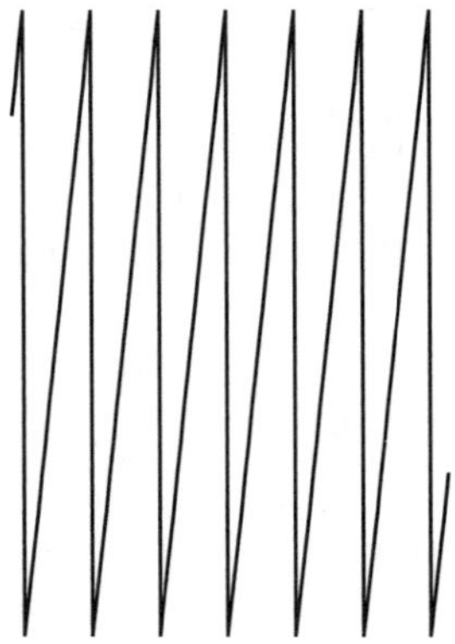

FIGURE 5.8 THE SAW WAVE OUTPUT.

The DoubleDataProducer Interface

The *DoubleDataProducer* interface is a public interface that resides in the *lyon.audio* package. The *DoubleDataProducer* interface isolates the *OscopeFrame* class (see next section) from application-specific code needed to obtain an array of data.

Classes that implement the *DoubleDataProducer* interface must provide methods whose signature is consistent with lines 3 and 4:

```
1.      package lyon.audio;
2.      public interface DoubleDataProducer {
3.      double [] getDoubleData();
4.      void openDataFile();
5.      }
```

For an example of usage, see the following section.

The OscopeFrame Class

The *OscopeFrame* class is a public abstract class in the *lyon.audio* package. The *OscopeFrame* can be extended to add an oscilloscope-type feature to any frame. *OscopeFrame* has its own paint method, so the user must be careful to call *super.paint* if a paint method is implemented in the subclass.

OscopeFrame has four private scroll bars whose events are handled locally. Thus, any subframe that handles events must be sure to call *super.handleEvent* at the end of the *handleEvent* method. Also, the *OscopeFrame* has a main menu bar setting. It can be overridden by the subclass, although doing so will remove the main menu bar features.

OscopeFrame extends the *PictFrame* class, which implements a feature in the main menu bar that enables the *OscopeFrame* to save a copy of itself as a pict file. The vector part of the pict file format is used so that the *OscopeFrame* has a vector output that can be edited with any draw program (including Microsoft Word's draw utility). Figure 5.9 is an example of a vector output generated by *OscopeFrame*. An example of the main menu bar that the OscopeFrame sets up is shown in Figure 5.10.

FIGURE 5.9 AN EXAMPLE OF THE VECTOR OUTPUT FROM OSCOPEFRAME.

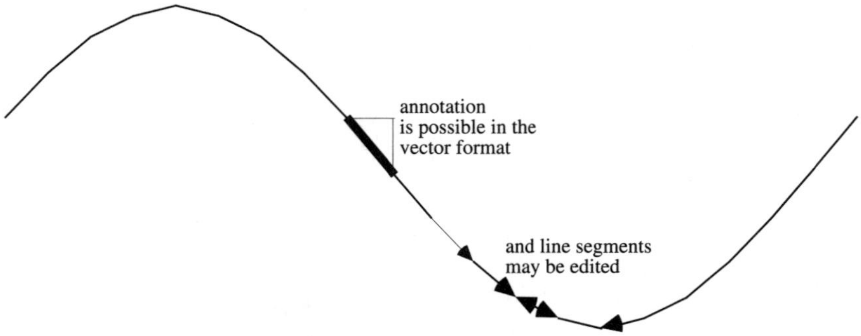

FIGURE 5.10 THE OSCOPEFRAME SAVE MENU PERMITS THE SAVING OF VECTOR OUTPUT.

OscopeFrame implements the *DoubleDataProducer* interface. As a result, the programmer must implement the *getDoubleData* and *openDataFile* methods, or the subclass will be abstract. When *openDataFile* is invoked, the subclass must respond by invoking *setDoubleData*.

An example of the *OscopeFrame* is shown in Figure 5.11.

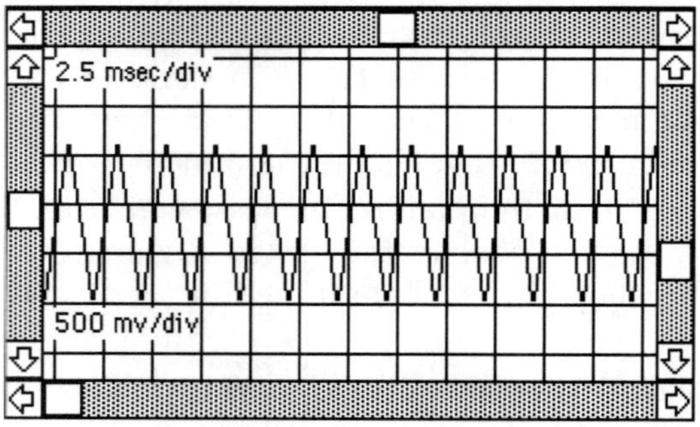

FIGURE 5.11 THE TRIANGLE WAVE IN OSCOPEFRAME.

Scroll bar adjustments permit translation and scaling of the waveform in both the time domain and the amplitude domain. Another view of the triangle wave of Figure 5.11 is shown in Figure 5.12.

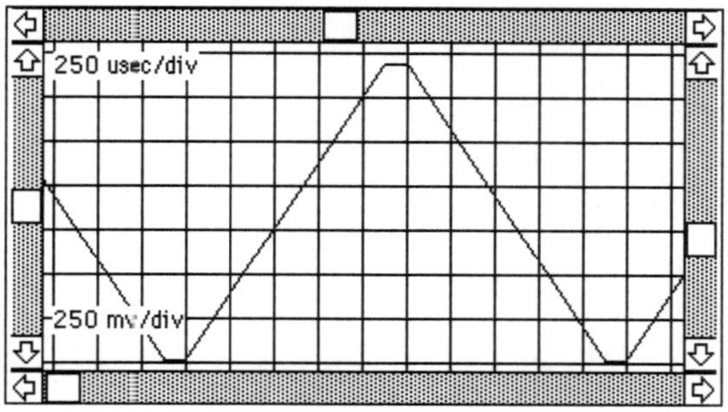

FIGURE 5.12 ANOTHER VIEW OF THE TRIANGLE WAVE IN THE OSCOPEFRAME.

From Figure 5.12 we can visually measure ten 250-microsecond divisions cycles in the waveform. This corresponds to a frequency of 1/(2,500 micro seconds) = 400 Hz. This is exactly what was specified for the oscillator (a multiple of the 8,000-Hz sampling rate).

Class Summary

```
package lyon.audio;
import java.awt.*;
import futils.utils.Draw;
import futils.Futil;
import lyon.dclap.PictFrame;


public abstract class OscopeFrame
     extends PictFrame implements DoubleDataProducer {
public OscopeFrame(String title)
public void setDoubleData(double [] d)
public void setGridColor(Color c)
public void paint(Graphics g)
public boolean handleEvent(Event e)
}
```

Class Usage

Because *OscopeFrame* implements the *DoubleDataProducer* interface, the subclass must implement the *getDoubleData* and *OpenDataFile* methods. The *AudioFrame* subclasses the *OscopeFrame*, as shown in the following code:

```
package lyon.audio;
import java.awt.*;
import gui.*;
import grapher.*;
import observers.*;
import futils.utils.*;
import futils.bench.*;
import futils.Futil;
import lyon.dclap.PictFrame;
public class AudioFrame extends OscopeFrame {
     UlawCodec ulc;
...
```

The *AudioFrame* constructor shows the name of the frame being passed to *OscopeFrame*:

```
public AudioFrame(String name) {
          super(name);
....
```

The *getDoubleData* and *OpenDataFile* methods are defined to keep the *AudioFrame* class from being abstract:

```
public double [] getDoubleData() {
          return ulc.getDoubleArray();
     }
public void openDataFile() {
     ulc = new UlawCodec();
     setDoubleData(ulc.getDoubleArray());
     ulc.playAsync();
}
```

Although no paint method was defined in *AudioFrame*, there were events. Thus, at the end of the event handler we are careful to call *super.handleEvent* so that the *OscopeFrame* events are processed:

```
public boolean handleEvent(Event e) {
     if (Evt.match(e,saveSound_mi)) {
          saveAs();
          return true;
     }....
          return super.handleEvent(e);
     }
```

Modifying OscopeFrame

Once a user starts to use the *OscopeFrame* class, there may be a need to add features without altering the existing API. One approach is to add to the existing menu bar. In the following code, we demonstrate how to add the capability to turn the *OscopeFrame* grid on and off using a menu item in the main menu bar. *OscopeFrame* extends *PictFrame*, and, inheriting an existing menu bar. To add an item to the menu bar, we add the following instance variables to the *OscopeFrame* class:

```
private boolean drawGrid = true;
private MenuBar mb = getMenuBar();
private Menu m = new Menu("Scope Menu");


private MenuItem drawGrid_mi =
     new MenuItem("[d]raw grid");
```

When the user presses **d** or selects the **[d]raw grid** item from the main menu bar, we want to turn off the drawing of the grid. The first step is to modify the *handleEvent* method:

```
public boolean handleEvent(Event e) {
     if(Evt.match(e,drawGrid_mi)) {
          drawGrid = ! drawGrid;
          return true;
     }....
```

Second, we modify is in the *drawTrace* method:

```
private void drawTrace(Graphics g) {
     doubleData = getDoubleData();
```

```
double limit = (double)(doubleData.length * xScaleFactor);


dim = size();
     height = dim.height;
     width = dim.width;
```

In the following line, we check the status of the *drawGrid* variable to see whether a grid should be drawn.

```
     if (drawGrid)
          Draw.grid(20, g, gridColor);
....
```

Switching the Scaling Labels

A unique feature of *OscopeFrame* is that it automatically switches to a next logical range when the user clicks in the time-scale or voltage-scale scroll bar. This interface hack is clever enough to warrant some explanation.

OscopeFrame has a series of predefined scale factors, with corresponding labels, for time scaling (y-scale) and voltage scaling (x-scale). They are held as private class variables:

```
1.      private final double
2.      xScaleFactors[] = {50000, 25000, 10000, 5000,  2500, 1000,
3.    250,    100,    50,     25,    10,
4.    2.5,      1,    .5,    .25,    .1,
5.   .025,    .01,  .005,  .0025,  .001,
6.;

7.      private final String xSFLabels[] = {"0.05 u", "0.1 u", "0.25 u",
8.                                          "0.5 u",   "1 u",  "2.5 u",
9.                                               "5 u",  "10 u",   "25 u",
10.                                               "50 u", "100 u",  "250 u",
11.                                               "500 u",   "1 m",  "2.5 m",
12.                                               "5 m",  "10 m",   "25 m",
13.                                               "50 m", "100 m",  "250 m",
14.                                               "500 m",    "1 ",   "2.5 ",
15.                                               "5 "};
```

NOTE

xsfStartIndex indicates the default start point for the scale factor and its label. This has a direct correspondence with the y-scale factors, shown in line 32.

```
    private final int xsfStartIndex = 14;
    private double xScaleFactor = xScaleFactors[xsfStartIndex];
    private double oldXScaleFactor = xScaleFactors[xsfStartIndex];
    private String xSFLabel = new String(xSFLabels[xsfStartIndex]);

    private final double
    yScaleFactors[] = {
      200, 100, 50, 20, 10,
        2,   1, .5, .2, .1};

    private final String
        ySFLabels[] = {
            "1 m", "2.5 m",
            "5 m", "10 m",
            "25 m", "50 m",
            "100 m", "250 m",
            "500 m",   "1 ",
            "2.5 ",   "5 "};

    private final int ysfStartIndex = 8;
    private double yScaleFactor = 1;
    private double oldYScaleFactor = 1;
    private String ySFLabel = new String(ySFLabels[ysfStartIndex]);
```

The clever bit is that the top and bottom scroll bar instances return integers that can be used to index into the label and scale arrays. This technique is shown in the following code, which is excerpted from the *handleEvent* method:

```
if(e.target == sbHorzTop) {
     int i = sbHorzTop.getValue();
     xScaleFactor = xScaleFactors[i];


     if (xScaleFactor != oldXScaleFactor) {
          xSFLabel = xSFLabels[i];
          repaint();
          oldXScaleFactor = xScaleFactor;
     }
     return true;
}
```

The scroll bar characteristics are set so that they will return only a valid range for an index into the scale factors and scale labels:

```
// Set top scroll bar characteristics
sbHorzTop.setValues(xsfStartIndex, 0, 0, 24);
// Set right scroll bar characteristics
sbVertRight.setValues(ysfStartIndex, 0, 0, 11);
```

The DoubleGraph Class

Some programmers will not want to bother extending the *OscopeFrame* class. They may find the interface difficult to implement properly, or they may wish for a simpler API, which, although less general, would make the code easier to write and understand. The *DoubleGraph* class will make an *OscopeFrame* and handle all the scroll bar events and interface implementations for you.

The *DoubleGraph* class resides in the *lyon.audio* package but is so generally useful that it could appear in a utilities class.

Class Summary

```
public class DoubleGraph extends OscopeFrame {
     DoubleGraph(double d[],String title)
     public double[] getDoubleData()
     public void openDataFile()
}
```

Class Usage

The *AudioFrame* class has a public method called *graphSound*. It shows the typical usage for the *DoubleGraph* class instance:

```
public void graphSound() {
     new DoubleGraph(ulc.getDoubleArray(),"Sound");
}
```

The act of instantiation displays the *OscopeFrame*, with scroll bars.

NOTE

To avoid added complexity, we have not included features for altering an existing *DoubleGraph* frame. Just make another one. The old frame will be collected by the garbage collector when all references to it are removed. Note that the *DoubleGraph* instance has a life that ends after the *graphSound* method returns. The *DoubleGraph* instance will thus be a temporary one. It would be hard to build a simpler API for graphing an array of doubles.

Summary

This chapter has covered a lot of ground, much of it quickly and without rigor. Our basic objective is to give you just enough background to understand the basic results of Fourier analysis so that the code in Chapter 6 will make more sense.

A great deal of complexity is hidden from the user in the classes presented in this chapter. Code is reproduced only where appropriate to the exposition of the book. To get a close look at the code, you should study the programs on the CD-ROM.

Adding features to the classes presented in this chapter can be a fruitful exercise. For example, turning off the grid is just one feature that can be added to *OscopeFrame*. There are many others: coloring a grid, extending *OscopeFrame* to make a *DualTraceOscopeFrame*, creating an x vs. y plot in *OscopeFrame*, and so on.

Chapter 6

Digital Audio Transform Recipes

> ...then thus he cried,
> When I the process have in memory,
> How thou hast wearied me on every side...
>
> —*William Wordsworth, 1888*

This chapter covers the finer points of one-dimensional digital signal processing. Chapter 5 showed some of the mathematical basis of the Fourier transform and pointed out some of its properties. This chapter explains how to implement the Fourier transform using uniformly sampled audio. We start with a description of the DFT (discrete Fourier transform) and IDFT (inverse discrete Fourier transform). We show how you can compute the DFT and IDFT in Java and demonstrate why such operations are usually slow. We discuss how to speed the DFT and IDFT using a class of algorithms known as the FFT (fast Fourier transform) and the IFFT (inverse fast Fourier transform).

We also demonstrate the computation of the PSD (power spectral density) mentioned in chapter 5. The remaining portion of the chapter is dedicated to applications of these transforms. The applications include filtering, windowing, pitch-shifting, and the spectral analysis of resampling.

The Discrete Fourier Transform

Let

$$v_j, j \in [0 \ldots N-1] \tag{6.1}$$

be the sampled version of the waveform $v(t)$, where N is the number of samples.

NOTE: Equation 6.1 numbers from zero, rather than 1, to reflect the start point of arrays in Java.

Recall that the Fourier transform of *v(t)* was given by

$$V(f) = F[v(t)] = \int_{-\infty}^{\infty} v(t)e^{-2\pi ift}\,dt \qquad (6.2)$$

NOTE: V(f) can exist only if *v*(*t*) is absolutely integrable:

$$\int_{-\infty}^{\infty} |v(t)|\,dt < +\infty$$

Recall also that the inverse Fourier transform of *V(f)* was given by

$$v(t) = F^{-1}\left[V(f)\right] = \int_{-\infty}^{\infty} V(f)e^{2\pi ift}\,dt \qquad (6.3)$$

To compute Equation 6.2 for the sampled waveform of Equation 6.1, we must perform DFT, which is given by

$$V_k = \frac{1}{N}\sum_{j=0}^{N-1} e^{-2\pi ijk/N} v_j \qquad (6.4)$$

Recall, from Chapter 5, Euler's relation:

$$e^{i\theta} = \cos\theta + i\sin\theta \qquad (6.4a)$$

This permits the alternative formulation of the kernel of the transform in Equation 6.4b:

$$e^{-2\pi ijk/N} = \cos(-2\pi jk/N) + i\sin(-2\pi jk/N) \tag{6.4b}$$

Recall that an even function is one that has the property $f(-x) = f(x)$. Also, an odd function has the property $f(-x) = -f(x)$. Because sin θ = -sin θ sine is an odd function. Also, because cos θ = -cos θ cosine is an even function. Using the odd-even function properties of sine and cosine, we rewrite Equation 6.4b:

$$e^{-2\pi ijk/N} = \cos(2\pi jk/N) - i\sin(2\pi jk/N) \tag{6.4c}$$

Substituting Equation 6.4c into Equation 6.4, we obtain the form of the DFT used in the Java implementation:

$$V_k = \frac{1}{N}\sum_{j=0}^{N-1}\left(\cos(2\pi jk/N) - i\sin(2\pi jk/N)\right)v_j \tag{6.5}$$

When implementing the Java program, we compute the real and imaginary parts of Equation 6.5 using the following:

```
for(int j = 0; j < N; j++) {
   twoPijkOnN = twoPikOnN * j;
   r_data[k] += v[j] * Math.cos( twoPijkOnN );
   i_data[k] -= v[j] * Math.sin( twoPijkOnN );
}
r_data[k] /= N;
i_data[k] /= N;
```

The complete routine for the DFT is given at the end of this section. Recall that the preceding loop is executed N times. The index, k , varies $0\ldots N{-}1$, because the DFT divides the spectrum into N parts. Each spectral division is accessed using the *frequency index*, k. The frequency, f_k, is computed from k using

$$f_k = \frac{k}{N\Delta t} \tag{6.6}$$

where the sampling period, Δ, is given by

$$\Delta t = 1/f_s \tag{6.7}$$

An examination of Equations 6.5 and 6.6 shows that the spectrum is described by integral harmonics of f_s / N. For example, suppose that the sampling rate, f_s, is 8,000 Hz and that the number of points is 2,048; then the smallest change in frequency that can be detected is given by

$$f_1 = \frac{1}{2048} 8000 \approx 4Hz$$

Therefore, integral harmonics of 4 Hz will be used to approximate *v(t)*.

The PSD (power spectral density) gives the power at a specific frequency index, psd_k. We compute the power at any f_k by summing the square of the real and imaginary components of the amplitude:

$$psd_k = real^2(V_k) + imaginary^2(V_k) \tag{6.8}$$

The amplitude spectral density is given by the square root of the power spectral density.

CD-ROM

An implementation of the DFT is shown next.

```
package lyon.audio;
import java.io.*;
import java.awt.*;
import grapher.Graph;
public class FFT extends Frame {
 double r_data[] = null;
 double i_data[] = null;
 ...
  public double[] dft(double v[]) {
   int N = v.length;


double t_img, t_real;


double twoPikOnN;
   double twoPijkOnN;
```

```
// how many bits do we need?
   N=log2(N);
   //Truncate input data to a power of two
   // length = 2**(number of bits).
    N = 1<<N;

double twoPiOnN = 2 * Math.PI / N;
   // We truncate to a power of two so that
   // we can compare execution times with the FFT.
   // DFT generally does not need to truncate its input.

r_data = new double [N];
   i_data = new double [N];
   double psd[] = new double[N];

System.out.println("Executirg DFT on "+N+" points...");

for(int k=0; k<N; k++) {

twoPikOnN = twoPiOnN *k;

for(int j = 0; j < N; j++) {
            twoPijkOnN = twoPikOnN * j;
         r_data[k] += v[j] * Math.cos( twoPijkOnN );
         i_data[k] -= v[j] * Math.sin( twoPijkOnN );
      }
      r_data[k] /= N;
      i_data[k] /= N;

psd[k] =
           r_data[k] * r_data[k] +
           i_data[k] * i_data[k];

}
   return(psd);
 }
```

Bit Computations and a Log Review

We have noticed that some students are confused by the use of logs in the following code:

```
1.    // how many bits do we need?
2.    N=log2(N);
3.    //Truncate input data to a power of two
4.    // length = 2**(number of bits).
5.    N = 1<<N;
```

If lines 2 and 5 are obvious to you, you should probably skip to the next section. Otherwise, please read on! Logs are basic things, but people seem to forget them.

The log function is defined by

$$\text{if } x = a^y \text{ then } y = \log_a x$$

For example, if $x = 210$, then $10 = \log_2 x$.
The following are known as the laws of logarithms:

$$\log_a(xy) = \log_a x + \log_a y$$

For example, to find $\log_2 4096$ use the following:

$$4096 = 2^y$$

$$\ln 4096 = y \ln 2$$

$$\frac{\ln 4096}{\ln 2} = y = 12$$

Where ln is the natural log. Therefore,

$$\log_2 x = \frac{\ln x}{\ln 2}$$

Also, to find $\log B_x$ use this:

$$\log_B x = \frac{\ln x}{\ln B} = \frac{\log_{10} x}{\log_{10} B} \qquad (6.9)$$

Based on Equation 6.9, line 2 computes the number of bits needed, rounding down, via truncation, the integral number, *N*. To compute Equation 6.9 to the base 2, we use the following:

```
public static int log2(int n) {
     return (int) (Math.log(n)/Math.log(2.0));
}
```

Thus, line 2 replaces *N*, the number of items in the list plus 1, with the log of *N* to the base 2, rounded *down* (also known as the *floor function*) to the nearest integer.

```
N=log2(N);
```

Finally, line 5 creates a number that is an integral power of two:

```
N = 1<<N;
```

NOTE

DFT does not need to truncate the length of the data to be a power of two. We will see, however, that there are implementations of the FFT that require this truncation. To compare the performance of the DFT implementation fairly against the FFT implementation, we must perform the same truncation for both algorithms.

The futils.Timer Class: Benchmarking the DFT

Benchmarking is a craft that measures hardware and software performance. One of the remarkable things about Java is that the compile-once-run-anywhere attribute enables the same byte codes to be executed on different implementations of the Java machine. In addition, we have similar Java virtual machines that are implemented on different hardware platforms. This arrangement enables a baseline comparison of different hardware platforms when executing the byte codes. Benchmarking can measure the effect of the implementation of an algorithm on execution time. Also of interest is the execution time of two different algorithms for the same data.

NOTE

There is a difference between the performance measurement of various algorithmic *implementations* and the performance measurement of various *algorithms*. Better algorithms are generally combined with better implementations to yield performance improvement

Class Summary

```
package futils.bench;
import java.io.*;
public class Timer {
     public Timer()
     public void mark()
     public void record()
     public float elapsed()
     public void report(PrintStream pstream)
     public void report()
}
```

Class Usage

Suppose the following variables are predefined:

```
Timer t1;
PrintStream pstream;
```

To make an instance of a *Timer* class (with the elapsed time set to zero), use the following:

```
t1 = new Timer();
```

To mark the time (similar to resetting a stopwatch):

```
t1.mark();
```

To add the elapsed time to the total elapsed time since the *record* method was last invoked:

```
t1.record();
```

To return the elapsed time in seconds:

```
t1.elapsed();
```

To print the elapsed time, in seconds, to the *PrintStream* instance *pstream*:

```
t1.report(pstream);
```

To print the elapsed time, in seconds, to *System.out*:

```
t1.report();
```

Benchmarking the DFT Method

In this section we show how to use the *Timer* class to measure the execution time of the DFT. This example shows how to use the DFT and how to benchmark a method. Benchmarking showed that executing the DFT on 2,048 points took 55 seconds on a 100-MHz PPC 601 without a JIT compilier. The following example of how to use the DFT in the *lyon.audio* package is contained in the *AudioFrame* class:

```
public void dft() {

     FFT f = new FFT();
     double [] doubleData = ulc.getDoubleArray();
     double [] psd;


      // Time the fft
      Timer t1 = new Timer();
      t1.mark();


     psd=f.dft(doubleData);


      // Stop the timer and report.
      t1.record();
      System.out.println("Time to perform DFT:");
      t1.report();
     f.graphs();
     new DoubleGraph(psd,"psd");
}
```

The Inverse DFT

Recall that the inverse Fourier transform of *V(f)* was given by

$$v(t) = F^{-1}[V(f)] = \int_{-\infty}^{\infty} V(f)e^{2\pi i f t}dt$$

To solve this for the sampled waveform (given by Equation 6.1), we must perform an IDFT (Inverse Discrete Fourier Transform which is given by:

$$V_k = \sum_{j=0}^{N-1} e^{2\pi ijk/N} v_j \qquad (6.10)$$

NOTE

Recall Equation 6.4:

$$V_k = \frac{1}{N} \sum_{j=0}^{N-1} e^{-2\pi ijk/N} v_j$$

The summation result is multiplied by $1/N$. This is not the case in Equation 6.10. In some expositions, both Equations 6.10 and 6.4 are multiplied by

$$1/\sqrt{N}$$

to keep the DFT and IDFT symmetric. We abandoned such an approach during development, because it complicates the presentation of the PSD and requires slightly more computation.

Recall, from Chapter 5, Euler's relation:

$$e^{i\theta} = \cos\theta + i\sin\theta$$

Substituting Equation 6.4a into Equation 6.10 results in

$$v_k = \sum_{j=0}^{N-1} \left[\cos(2\pi jk/N) + i\sin(2\pi jk/N)\right] V_j \qquad (6.11)$$

Recall that when two complex numbers are multiplied together, $z_1 z_2$, the result can be expressed as

$$z_1 z_2 = (x_1 + iy_1)(x_2 + iy_2) = x_1 x_2 - y_1 y_2 + i(x_1 y_2 + y_1 x_2) \qquad (6.11a)$$

Based on Equation 6.11a, we conclude that the real part of $z_1 z_2$ is given by

$$real(z_1 z_2) = x_1 x_2 - y_1 y_2 \qquad (6.11b)$$

and the imaginary part of z_1z_2 is given by

$$imaginary(z_1z_2) = x_1y_2 + y_1x_2 \tag{6.11c}$$

We are interested in a reconstruction of a real signal, so we do not need to compute the result in Equation 6.11c. Substituting Equation 6.11b into Equation 6.11 yields

$$real(v_k) = \sum_{j=0}^{N-1}\left[\cos(2\pi jk/N)real(V_j) - \sin(2\pi jk/N)imaginary(V_j)\right] \tag{6.12}$$

Computing only the real part of the IDFT saves $2N$ multiplications and N additions for each frequency index, k. A comparison of Equation 6.12 with Equation 6.5,

$$V_k = \frac{1}{N}\sum_{j=0}^{N-1}\left(\cos(2\pi jk/N) - i\sin(2\pi jk/N)\right)v_j$$

The following code implements the IDFT shown in Equation 6.12:

```
   // assume that r_data and i_data are
   // set. Also assume that the real
   // value is to be returned
public double[] idft() {
    int N = r_data.length;
    double twoPiOnN = 2 * Math.PI / N;

    double twoPikOnN;
    double twoPijkOnN;

    double v[] = new double[N];

    System.out.println("Executing IDFT on "+N+" points...");

    for(int k=0; k<N; k++) {
       twoPikOnN = twoPiOnN *k;
       for(int j = 0; j < N; j++) {
          twoPijkOnN = twoPikOnN * j;
          v[k] +=  r_data[j] * Math.cos(twoPijkOnN )
                   - i_data[j] * Math.sin(twoPijkOnN );

       }
   }
 return(v);
}
```

Numeric Check of the DFT and IDFT

A numeric check should be an integral part of every class. The *FFT* class contains a method called *testDFT* whose sole job is to verify the correctness of the DFT and IDFT implementation. With the number of samples set to 8, the testing method prints the input and output points for comparison.

```
public static void testDFT() {

    int N = 8;
     FFT f = new FFT(N);


    double v[];


    double x1[] = new double[N];
     for (int j=0; j<N; j++)
          x1[j] = j;

    // take dft
     f.dft(x1);


    v = f.idft();
     System.out.println("j\tx1[j]\tre[j]\tim[j]\t v[j]");
     for (int j=0; j < N; j++)
         System.out.println(
         j+"\t"+
         x1[j]+"\t"+
         f.r_data[j]+"\t"+
         f.i_data[j]+"\t"+
         v[j]);


}
```

NOTE

We print the intermediate DFT results to permit a detailed check against variations in implementation. Full data disclosure allows a baseline comparison among different implementations of the DFT. As we shall see in the following section, this data result is comforting, particularly when compared with the FFT data.

```
Executing IDFT on 8 points...
j    x1[j]    re[j]        im[j]        v[j]
0    0        3.5          0            -3.10862e-15
1    1        -0.5         1.20711      1
2    2        -0.5         0.5          2.00000
3    3        -0.5         0.207107     3
4    4        -0.5         0            4
5    5        -0.500000    -0.207107    5
6    6        -0.500000    -0.5         6
7    7        -0.5         -1.20711     7
```

Although the input is not quite the same as the output, it is quite close.

The FFT

The FFT is a family of algorithms designed to speed the computation of the DFT:

$$V_k = \frac{1}{N}\sum_{j=0}^{N-1} e^{-2\pi ijk/N} v_j$$

Direct computation of the DFT takes $O(N^2)$ complex multiplications whereas the FFT takes $O(N \log N)$ complex multiplications. In this section we show an FFT algorithm known as the decimation in time, radix 2 FFT algorithm (also known as the Cooley-Tukey algorithm). The Cooley-Tukey algorithm is probably one of the most widely used of the FFT algorithms. The radix 2 property means that the number of samples to be transformed must be a power of two. The decimation in time means that the algorithm performs a recursive subdivision of the input sequence into its odd and even members. We are able to perform this subdivision as a result of the Danielson-Lanczos lemma:

$$V_k = \frac{1}{N}\left[V_k^e + W^k V_k^o\right] \qquad \forall_{k\in[0\ldots N-1]}$$

Here is the proof of the Danielson-Lanczos lemma:
Let

$$W = e^{-2\pi i/N} \tag{6.13}$$

so that

$$W^{jk} = W^{j}W^{j(k-1)}$$

Substitute Equation 6.13 into Equation 6.4 to obtain

$$V_k = \frac{1}{N}\sum_{j=0}^{N-1} W^{jk} v_j \tag{6.14}$$

Separate Equation 6.14 into its odd and even components by altering the way the samples are indexed.

$$V_k = \frac{1}{N}\left[\sum_{j=0}^{N/2-1} W^{2jk} v_{2j} + \sum_{j=0}^{N/2-1} W^{(2j+1)k} v_{2j+1}\right] \tag{6.15}$$

Where Equation 6.15 shows summations operating over the odd and even indices. For example, if

$$j = 0,1,2,3...$$

then

$$2j = 0,2,4,6... \text{ and } 2j+1 = 1,3,5...$$

Factoring the exponents in Equation 6.15 yields

$$V_k = \frac{1}{N}\left[\sum_{j=0}^{N/2-1} W^{2jk} v_{2j} + \sum_{j=0}^{N/2-1} W^{2jk} W^{k} v_{2j+1}\right]$$

The W^k term in the rightmost summation is not a function of the index, so

$$V_k = \frac{1}{N}\left[\sum_{j=0}^{N/2-1} W^{2jk} v_{2j} + W^{k} \sum_{j=0}^{N/2-1} W^{2jk} v_{2j+1}\right] \tag{6.16}$$

To reflect the odd and even summations, Equation 6.16 is rewritten as

$$V_k = \frac{1}{N}\left[V_k^e + W^k V_k^o\right] \qquad \forall_{k \in [0 \ldots N-1]} \tag{6.17}$$

Q.E.D.

The implications of Equation 6.17 are that we can divide the sequence into odd-and even-numbered samples. Thus, the Danielson-Lanczos lemma enables a divide and conquer algorithm to recursively split the sample sequence in half. The computational result of the Danielson-Lanczos lemma is that the $O(N^2)$ DFT can be computed in $O(N \log N)$ time.

The Danielson-Lanczos lemma shows that a sequence must be divided into its odd and even subsets and these subsets must in turn be divided into their subsets. This process continues until we have only two members per subset. An illustration of this subdivision, for N=*8*, is shown in Figure 6.1.

0-1-2-3-4-6-7

0-2-4-6 1-3-5-7

0-4 2-6 1-5 3-7

FIGURE 6.1 DECIMATION IN TIME.

It is natural to implement the decimation in time using recursive calls with odd and even sets. It has been shown, however, that a recursive implementation is six times slower than a nonrecursive implementation [Gonzalez et al.]. To avoid a recursive approach to decimation, the bit-reversal algorithm (Cooley-Tukey) is used to perform the decimation in time (see Figure 6.2).

N	A	B	C	C	B	A	bitr(N)
0	0	0	0	0	0	0	0
1	0	0	1	1	0	0	4
2	0	1	0	0	1	0	2
3	0	1	1	1	1	0	6
4	1	0	0	0	0	1	1
5	1	0	1	1	0	1	5
6	1	1	0	0	1	1	3
7	1	1	1	1	1	1	7

FIGURE 6.2 AN EXAMPLE OF HOW TO DECIMATE BY BIT REVERSAL.

To arrive at the bit reversal, we implement a Java method in the *FFT* class:

```
int bitr(int j) {
     int ans = 0;
     for (int i = 0; i< nu; i++) {
          ans = (ans <<1) + (j&1);
          j = j>>1;
     }
     return ans;
}
```

The *bitr* method works by linking two software shift registers, as shown in Figure 6.3.

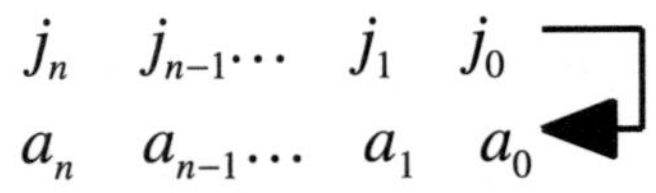

FIGURE 6.3 THE J AND A REGISTERS ARE LINKED WITH THE + OPERATOR.

After the decimation in time is performed, the balance of the computation is optimization hacks and housekeeping. For example, a simplification results from *Vk* being periodic in *N* so that $V_{k+N} = V_k$. Here is the proof:

Recall that the DFT is given by

$$V_k = \frac{1}{N}\sum_{j=0}^{N-1} e^{-2\pi ijk/N} v_j$$

so that

$$V_{k+N} = \frac{1}{N}\sum_{j=0}^{N-1} e^{-2\pi ij(k+N)/N} v_j$$

Expanding the exponents and simplifying using

$$V_{k+N} = \frac{1}{N}\sum_{j=0}^{N-1} e^{-2\pi ijk/N} e^{-2\pi ijN/N} v_j$$

with

$$e^{-2\pi ij} = \cos(-2\pi j) + i\sin(-2\pi j) = 1$$

Thus,

$$V_{k+N} = \frac{1}{N}\sum_{j=0}^{N-1} e^{-2\pi ijk/N} v_j$$

with

$$V_{k+N} = V_k \tag{6.18}$$

Q.E.D.
In addition, it can be shown that

$$W^{k+N/2} = -W^k \qquad 0 \le k \le N/2 \tag{6.19}$$

Here is the proof:
Recall Equation 6.13 is given by

$$W = e^{-2\pi i/N}$$

So that

$$e^{-2\pi i(k+N/2)/N} = \cos(-2\pi(k+N/2)/N) + i\sin(-2\pi(k+N/2)/N)$$

with

$$\begin{aligned} \cos(-2\pi(k+N/2)/N) &= \cos(2\pi k/N + \pi) = -\cos(2\pi k/N) \\ \sin(-2\pi(k+N/2)/N) &= \sin(2\pi k/N + \pi) = -\sin(2\pi k/N) \end{aligned} \tag{6.20}$$

This naturally leads to

$$W^{k+N/2} = -W^k \qquad 0 \le k \le N/2$$

Q.E.D.
A further efficiency can be gained by the use of the recurrence relation:

$$W^{jk} = W^j W^{j(k-1)} \tag{6.21}$$

Here is the proof:

$$W^{jk} = e^{-2\pi ijk/N} = \cos\left(-2\pi jk/N\right) + i\sin(-2\pi jk/N)$$

$$W^{jk} = \cos\left(-2\pi jk/N\right) + i\sin(-2\pi jk/N)$$

$$W^{jk} = \left[\cos\left(-2\pi j/N\right) + i\sin(-2\pi j/N)\right] * \left[\cos\left(-2\pi j(k-1)/N\right) + i\sin(-2\pi j(k-1)/N)\right]$$

$$W^{jk} = W^{j} W^{j(k-1)} \tag{6.22}$$

Q.E.D.

The real part of Equation 6.22 is given by

$$real(z_1 z_2) = x_1 x_2 - y_1 y_2$$

so that

$$W_r^{jk} = W_r^j W_r^{j(k-1)} - W_i^j W_i^{j(k-1)} \tag{6.23}$$

The imaginary part of Equation 6.22 is given by

$$imaginary(z_1 z_2) = x_1 y_2 + y_1 x_2$$

so that

$$W_i^{jk} = W_r^j W_i^{j(k-1)} + W_i^j W_r^{j(k-1)} \tag{6.24}$$

Equations 6.23 and 6.24 form the basis of the recurrence relationships, which enables the quick computation of the next W^{jk} based on the previous *Wjk*. An implementation of Equation 6.24 follows:

```
1.          // (eq 6.23) and (eq 6.24)
2.          wtemp = Wjk_r;
3.          Wjk_r = Wj_r * Wjk_r  - Wj_i * Wjk_i;
4.          Wjk_i = Wj_r * Wjk_i  + Wj_i * wtemp;
```

Line 2 introduces *wtemp*, a temporary variable that enables the multiplication of the two complex numbers.

The FFT Class

The *FFT* class is a public class that resides in the *lyon.audio* package. It depends on the *grapher.Graph* package for performing graph functions, as well as the *futils.bench.Timer* class for benchmarking. If these packages are not needed, they can be removed along with their invocations. Such a situation might arise if this code were used in another program.

The *grapher* package provides a simple interface to make an automatically scaled graph. Usually, only a single method is invoked.

```
public void makeHann() {
double window[];
     window = makeHann(256);
     Graph.graph(window,
                "The Hann window","f");
}
```

The "The Hann window" string appears along the x-axis, and "f" appears on the y-axis. *Graph.graph* can be invoked directly, because the *graph* method is static. It graphs an array of type *double*.

Class Summary

```
package lyon.audio;
import java.io.*;
import java.awt.*;
import grapher.Graph;
import futils.bench.Timer;
public class FFT extends Frame {
  public FFT(int N)
  public FFT()
  public void graphs()
  public void graphs(String t)
  public void setTitle(String t)
  public static double getMaxValue(double in[])
  public static int log2(int n)
  public static double[] arrayCopy( double [] in)
  public double [] computePSD ()
  public double[] dft(double v[])
  public double[] idft()
  public double [] getReal()
```

```
  public double [] getImaginary()
  public void forwardFFT(double in_r[], double in_i[])
  public void reverseFFT(double in_r[], double in_i[])
  public void printArray(double[] v,String title)
  public void printArrays(String title)
  public void printReal(String title)
  public static void main(String args[])
  public static void timeFFT()
  public static void testFFT()
  public static void testDFT()
}
```

Class Usage

The *FFT* class maintains internal data arrays that are stored as doubles. These arrays are private and are used to assist computation. Furthermore, the in-place Cooley-Tukey algorithm employed for the fast transform is destructive of the original data. The *FFT* class in the *lyon.audio* package uses doubles for all computations. This class is for 1-D (audio) transforms.

Suppose the following variables are predefined:

```
FFT f;
int N = 8;
double inputArray[];
String title = "My data title";
double aDoubleArray[];
double in_r[];
double in_i[];
```

To make a new instance of the *FFT* class and allocate two internal arrays of double, each of length *N*, use:

```
f = new FFT(N);
```

To make a new instance of the *FFT* class with no memory allocation:

```
f = new FFT();
```

To graph the real and imaginary data arrays:

```
f.graphs();
```

To graph the real and imaginary data arrays with a title:

```
f.graphs(title);
```

To set the title for the graphs:

```
f.setTitle(title);
```

To get the maximum value of an *inputArray*:

```
FFT.getMaxValue(inputArray);
```

To compute the floor of the log of an *int* to base 2:

```
int numberOfBits = FFT.log2(N);
```

To copy an array of double:

```
aDoubleArray = FFT.arrayCopy(inputArray);
```

To compute the PSD (power spectral density) of the previous DFT or FFT:

```
aDoubleArray = f.computePSD();
```

To nondestructively compute the DFT of an input array and return the PSD:

```
aDoubleArray = f.dft(inputArray);
```

NOTE

DFT, IDFT, FFT, and IFFT alter the internal data structures in an instance of the *FFT* class.

To get the real part of the last transform:

```
aDoubleArray = f.getReal();
```

To get the imaginary part of the last transform:

```
aDoubleArray = f.getImaginary();
```

To take the IDFT of the internal data and return the real part:

```
aDoubleArray = f.idft();
```

To take the forward FFT on two input arrays, (destructively):

```
f.forwardFFT(in_r, in_i);
```

To take the inverse FFT on two input arrays, (destructively):

```
f.reverseFFT(in_r, in_i);
```

To print an array of double, with a title:

```
f.printArray(aDoubleArray, title);
```

To print the internal real and imaginary arrays, with a title:

```
f.printArrays(title);
```

To print the internal real array, with a title:

```
f.printReal(title);
```

To test the DFT, IDFT, FFT, and IFFT:

```
FFT.main();
```

To time the FFT:

```
FFT.timeFFT();
```

To test the FFT:

```
FFT.testFFT();
```

To test the DFT:

```
FFT.testDFT();
```

Testing the FFT and IFFT

The *FFT* class has a static method that permits the testing of the DFT, IDFT, FFT, and IFFT. It also performs timing for a transform of 2,048 doubles. To run this test, you must invoke the following:

```
FFT.main();
```

The code for the *FFT.main* method follows:

```
public static void main(String args[]) {
     testDFT();
     timeFFT();
     testFFT();
}
```

The test methods are run on an eight-point input array consisting of a linear ramp, providing a short sequence of input data that can be verified by printing. The timing is performed on 2,048 samples stored in two arrays of 2,048 doubles each (real and imaginary). The output of the main method follows:

```
Executing DFT on 8 points...
Executing IDFT on 8 points...
j     x1[j]     re[j]      im[j]      v[j]
0     0         3.5        0          -3.10862e-15
1     1         -0.5       1.20711    1
2     2         -0.5       0.5        2.00000
3     3         -0.5       0.207107   3
4     4         -0.5       0          4
5     5         -0.500000  -0.207107  5
6     6         -0.500000  -0.5       6
7     7         -0.5       -1.20711   7
fft: bit reversal
Time for 2048point fftTime 0.178000 sec
fft: bit reversal
Time for 2048point ifftTime 0.164000 sec
Starting 1D FFT test...
fft: bit reversal
fft: bit reversal
j     x1[j]     re[j]     im[j]      v[j]
0     0         3.5       0          0
1     1         -0.5      1.20711    1.00000
2     2         -0.5      0.5        2.00000
3     3         -0.500000 0.207107   3.00000
4     4         -0.5      0          4
5     5         -0.5      -0.207107  5
6     6         -0.5      -0.5       6
7     7         -0.500000 -1.20711   7
```

The input and output are highly correlated for both the DFT and FFT. The accuracy of these two radically different algorithms and implementations is surprising. Also, recall that the execution time for the DFT was benchmarked at 55 seconds. The FFT implementation is run in 0.178 seconds, a 308-fold improvement in speed. At 8,000 samples per second, the 2,048 samples represent 0.256 seconds of data on a limited data rate connection (such as a 28.8 kbps modem), the time to transmit the data is 2,048*8 bits /28,800 bits/sec = 0.56 seconds. Many dial-up users experience a slower connection than the maximum their modem permits. Thus, there is a window of opportunity for devising a real-time CODEC (In Java) that can perform FFT-based compression algorithms. An algorithm based on transform compression typically takes the original data, performs the forward transform, selects coefficients, quantizes, and then transmits. Data is recovered by performing an inverse transform on the coefficients. Very low bit rate voice compression (VLBRVC) is a rich and growing field that lies beyond the scope of this book. See http://www.bdti.com/faq/dsp_faq.htm for an FAQ that relates to this and other DSP topics.

Implementing the FFT.testFFT Method

The following code shows how to use the *FFT* class to perform a forward and inverse FFT. The static nature of the *testFFT* method indicates that you can invoke it without making an instance of the *FFT* class.

Line 3 makes an instance of the *FFT* class without allocating any memory for the internal data structures. Thus, arrays are allocated and copied outside the *forwardFFT* methods, in part, because of the destructive nature of the in-place Cooley-Tukey FFT algorithm. The trade-off is that the programmer must keep track of the data that is being processed by *forwardFFT*. The alternative is to automatically copy arrays, perform the in-place *forwardFFT*, and then return the copies. Our findings indicate that the dynamic allocation of memory (particularly during image processing, discussed later in this book) can slow performance as much as 100-fold. The housekeeping chores required of the programmer are warranted by the gain in performance.

```
1.     public static void testFFT()    {
2.     System.out.println("Starting 1D FFT test...");
3.     FFT f = new FFT();
```

Line 4 can be altered to any number of samples, *N*, but a large *N* will result in a large printout.

```
4.    int N = 8;
5.    int numBits = f.log2(N);
```

Lines 6–8 set up the input data as a ramp that varies from 0 to *N*.

```
6.    double x1[] = new double[N];
7.      for (int j=0; j<N; j++)
8.           x1[j] = j;
```

Now the housekeeping. To keep copies of the original data, the result of the forward FFT, and the result of the inverse FFT, you must allocate four arrays. This case is unusual, because it requires you to keep all intermediate results for checking purposes. Normally, production code would not have to keep all intermediate results.

```
9.    double[] in_r = new double[N];
10.    double[] in_i = new double[N];
```

The *in_r* and *in_i* arrays are copies of the input data, with the imaginary component equal to zero. Real data (such as audio data) often has a zero imaginary component, which lets some algorithms save significant time. This requires a different FFT implementation.

```
11.    double[] fftResult_r = new double[N];
12.    double[] fftResult_i = new double[N];
13.    // copy test signal.
14.    in_r = arrayCopy(x1);
```

Line 14 copies the input data into *in_r* and line 15 replaces *in_r* and *in_i* with the forward FFT results.

```
15.    f.forwardFFT(in_r, in_i);
16.    // Copy to new array because IFFT will
17.    // destroy the FFT results.
18.    fftResult_r = arrayCopy(in_r);
19.    fftResult_i = arrayCopy(in_i);
20.    f.reverseFFT(in_r, in_i);
21.    System.out.println("j\tx1[j]\tre[j]\tim[j]\tv[j]");
22.    for(int i=0; i<N; i++) {
23.        System.out.println(
24.            i + "\t" +
25.            x1[i] + "\t" +
26.            fftResult_r[i] + "\t" +
27.          fftResult_i[i] + "\t" +
28.          in_r[i]);
29.    }
30.     }
```

PSD Computations

To compute the PSD of the FFT output, we use the *computePSD* method in the *FFT* class. It squares the real and imaginary parts of the output of the FFT and places them in a new array.

```
public double [] computePSD () {
      double [] psd = new double[r_data.length];
      for (int k = 0; k < r_data.length; k++) {
        psd[k] =
           r_data[k] * r_data[k] +
           i_data[k] * i_data[k];

     }
      return psd;
 }
```

To test the PSD, we use a private method in the *FFT* class called *testPSD*.

```
private static void testPSD() {
    FFT f = new FFT();
    int N = 8;
    int numBits = f.log2(N);
    double x1[] = new double[N];
      for (int j=0; j<N; j++)
           x1[j] = j;
    double[] in_r = new double[N];
    double[] in_i = new double[N];
    // copy test signal.
    in_r = arrayCopy(x1);
    f.forwardFFT(in_r, in_i);
    f.printArrays("After the FFT");
    double psd[] = f.computePSD();
    FFT.printArray(psd,"The psd");
}
```

Here is the output of the test method.

```
After the FFT
[0]=(3.5,0)
[1]=(-0.5,1.20711)
[2]=(-0.5,0.5)
[3]=(-0.5,0.207107)
[4]=(-0.5,0)
[5]=(-0.500000,-0.207107)
[6]=(-0.500000,-0.5)
[7]=(-0.500000,-1.20711)
The psd
v[0]=12.25
v[1]=1.70711
v[2]=0.5
v[3]=0.292893
v[4]=0.25
v[5]=0.292893
v[6]=0.500000
v[7]=1.70711
Completed(0)
```

You can readily verify that the PSD is the sum of the squares of the real and the imaginary parts of the spectra (see Figure 6.14).

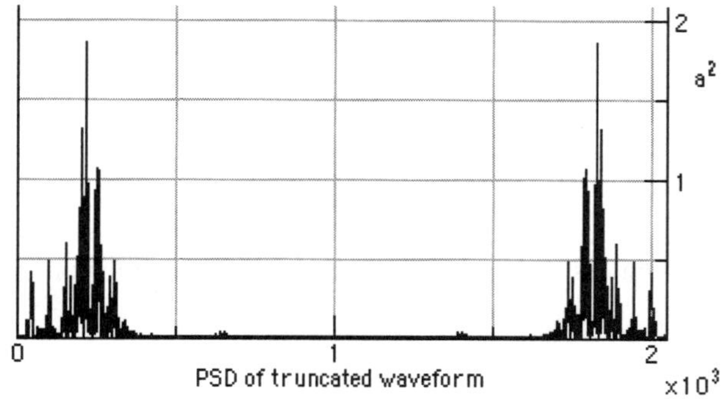

FIGURE 6.4 THE PSD OF A 2,048 SAMPLED WAVEFORM.

The implementation of the PSD computation and graphing is discussed next.

Implementation of the Transforms in the AudioFrame

In the *AudioFrame* class, we assume that we will be taking the FFT of a real signal. A real signal, such as audio, has no imaginary part; it has only a single value that varies from sample to sample. Thus, we construct a complex input to the FFT, setting the imaginary part of the input equal to zero.

When taking the IFFT, a signal that starts as real will end up as a real signal. The exception occurs when a spectral modification introduces terms that do not null out in the imaginary plane. This would happen, for example, if we added the spectra from real and imaginary signals.

Figure 6.5 shows the part of the Audio menu in the main menu bar in the *AudioFrame* that contains the transform fragments for performing the FFT manipulations. The keyboard shortcuts are shown in brackets.

[1] FFT
[2] IFFT
[3] DFT
[4] IDFT
[5] Graph PSD, R and I

FIGURE 6.5 THE TRANSFORM FRAGMENT OF THE AUDIOFRAME.

The menu items are initialized at the head of the *AudioFrame* class using the following code fragment:

```
MenuItem fft_mi        = addMenuItem(m,"[1] FFT");
MenuItem ifft_mi       = addMenuItem(m,"[2] IFFT");
MenuItem dft_mi        = addMenuItem(m,"[3] DFT");
MenuItem idft_mi       = addMenuItem(m,"[4] IDFT");
MenuItem graphPSD_mi   = addMenuItem(m,"[5] Graph PSD, R and I");
```

Events are handled in the *handleEvent* method, using the *Evt* class to detect the keyboard shortcuts. This technique enables rapid processing.

```
    if (Evt.match(e,fft_mi)) {
            fft();
            return true;
        }
    if (Evt.match(e,ifft_mi)) {
            ifft();
            return true;
        }


    if (Evt.match(e, dft_mi)) {
            fftInstance = dft();
            return true;
        }
    if (Evt.match(e, idft_mi)) {
            idft(fftInstance);
            return true;
        }


    if (Evt.match(e, graphPSD_mi)) {
        graphPSD();
        return true;
    }
```

A class variable called *fftInstance* passes the complex data results from method to method. The *fftInstance* is declared after the *Oscillator* and *UlawCodec* instances.

```
public class AudicFrame extends OscopeFrame {


    private     Cscillator osc =
        new Oscillator(400,4000);

    UlawCodec ulc;
    FFT fftInstance;
```

The following DFT and IDFT code includes facilities for benchmarking the execution of the DFT on long sequences. We find the DFT and IDFT to be so slow that they are of interest only for timing purposes. As a twist on the implementation possibilities, we also show the *dft* method returning an instance of the *FFT* class, which is used to replace the *fftInstance* class variable in the *handleEvent* method.

```
public FFT dft() {
     double doubleArray[] = ulc.getDoubleArray();
     double [] psd;
     FFT f = new FFT();
      Timer t1 = new Timer();
      t1.mark();
     psd=f.dft( doubleArray);
      // Stop the timer and report.
      t1.record();
      System.out.println("Time to perform DFT:");
      t1.report();



     f.graphs();
     return f;



}
```

The *handleEvent* method passes *fftInstance* to the *idft* method for processing. Thus, the *idft* method overwrites the *fftInstance* variable (it is passed, like all instances, by reference). The *idft* method follows:

```
public void idft(FFT f) {
     double doubleArray[];
      Timer t1 = new Timer();
      t1.mark();
     doubleArray=f.idft( );
      // Stop the timer and report.
      t1.record();
      System.out.println("Time to perform IDFT:");
      t1.report();
     f.graphs();
}
```

In the following *fft* method, *fftInstance* is overwritten. Data is obtained from the *getTruncatedDoubleData* method, which truncates the input data to the nearest integral power of two. We have elected to perform this truncation explicitly, because some users might prefer to pad their input data with extra samples. If the input data consisted of 2,049 samples, such a practice would pad the input data to 4,096 samples, doubling the processing time. The loss of data may be objectionable, so we have elected to perform the truncation, or padding, outside the *FFT* class.

WARNING

Be careful—if the input to the FFT is not an integral power of two, the *forwardFFT* and *reverseFFT* methods will produce wrong results, because no check is performed.

```
public void fft() {

        fftInstance = new FFT();
        double[] r_d = getTruncatedDoubleData();
        double[] i_d = new double[r_d.length];
        fftInstance.forwardFFT(r_d, i_d);
        ulc.play();

}
```

NOTE

The *UlawCodec* instance, *ulc*, is played after the *fft*.

The audio data *AudioFrame* class is not shortened or altered by the *fft* invocation. The play invocation simply signals when *fft* is complete.

The audio data is copied before truncation occurs. It is not overwritten until after the *ifft* method is invoked. The *getTruncatedDoubleData* truncation method follows:

```
double [] getTruncatedDoubleData() {
        double[] temp = FFT.arrayCopy(getDoubleData());

        int trunc = 1 << FFT.log2(temp.length);
        System.out.println("Truncated size: " + trunc);
        double[] truncArray = new double[trunc];
```

```
        for (int i=0; i < trunc; i++)
             truncArray[i] = temp[i];


        return truncArray;


}
```

Performing an FFT is sometimes called *analysis*. After the analysis, there may be a frequency-domain-based pattern recognition (what note was played) or a spectral modification (filtering, pitch-shifting, and so on). The IFFT is sometimes called *synthesis*, because the time-domain waveform is resynthesized from the frequency domain description. The implementation of the *ifft* method in the *AudioFrame* follows:

```
public void ifft() {
     double realInput[] =
          FFT.arrayCopy(fftInstance.getReal());
     double imaginaryInput[] =
          FFT.arrayCopy(fftInstance.getImaginary());
     fftInstance.reverseFFT(realInput,imaginaryInput);
     ulc = new UlawCodec(realInput);
     ulc.play();
}
```

NOTE

The *ifft* method throws away the imaginary part of the *reverseFFT* method's output. For real input signals, in theory the imaginary part is zero. We have found, however, that there is some small near-zero imaginary part. It has been attributed to round-off error.

The PSD is graphed using *graphPSD*:

```
public void graphPSD() {
             Graph.graph(fftInstance.computePSD(),
                  "PSD of truncated waveform","a^2");
     };
```

For example, suppose that a saw wave is used as an input. The saw wave and PSD are shown using the *OscopeFrame* and *graph* methods on the left and right in Figure 6.6.

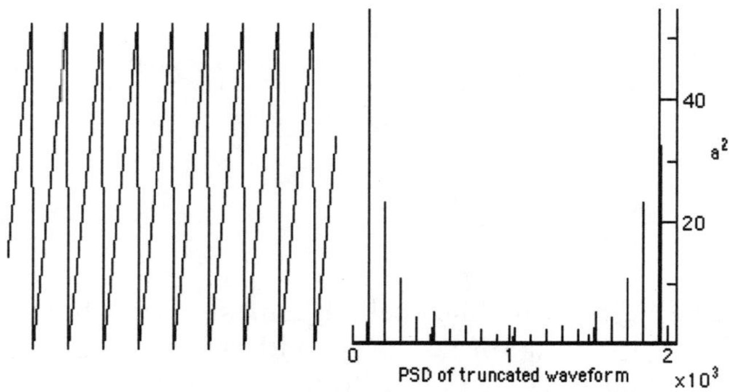

FIGURE 6.6 SAW WAVE AND SPECTRAL OUTPUT FROM THE GRAPHPSD METHOD.

NOTE

Real signals always have a spectrum that is symmetric about its center.

A Noise Filter Using the FFT

The basic idea of providing a noise filter is that you begin with a signal, add noise, perform an FFT on the signal, remove all spectral harmonics that have a PSD below some threshold, and then take the IFFT. Selecting the PSD threshold for noise can be tricky. What works well on a synthetic sound might turn a sampled sound into silence.

There are limits to the amount of noise that can be filtered out. A block diagram of the process appears in Figure 6.7. The following code adds noise to the waveform stored in the *AudioFrame* instance.

```
public void addNoise() {
     double r_d[] = getDoubleData();
     for (int i = 0; i< r_d.length; i++)
          r_d[i] = r_d[i] + 0.1*(Math.random() -0.5);
     ulc = new UlawCodec(r_d);
}
```

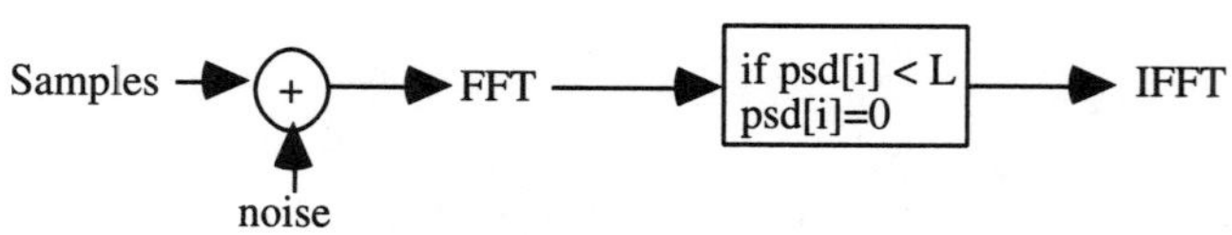

FIGURE 6.7 THE NOISE FILTER.

The *Math.random* method returns a random value between zero and 1. Thus, the sampled data is summed with time-domain uniformly distributed noise (also known as *white noise*) that varies from –0.10 to 0.10. The following code performs a PSD-based cutoff after taking the FFT of the sound samples:

```
public void removeNoise() {
     double noisePowerCutoff = 0.05;


        fftInstance = new FFT();
        double r_d[] = getTruncatedDoubleData();
        double i_d[] = new double[r_d.length];
        fftInstance.forwardFFT(r_d, i_d);
        double psd[] = fftInstance.computePSD();
        for (int i = 0; i < psd.length; i++) {
             if (psd[i] < noisePowerCutoff) {
                  r_d[i] = 0;
                  i_d[i] = 0;
             }
        }
        fftInstance.reverseFFT(r_d,i_d);
          ulc = new UlawCodec(r_d);
          ulc.play();
}
```

The *removeNoise* and *addNoise* methods are placed under interactive user control via the main menu bar entries:

```
MenuItem addNoise_mi      = addMenuItem(m,"[A] Add Noise");
MenuItem removeNoise_mi      = addMenuItem(m,"[R] Remove Noise");
```

The initial waveform is a sine wave of 400 Hz. A graph of the sine wave with PSD is shown in Figure 6.8.

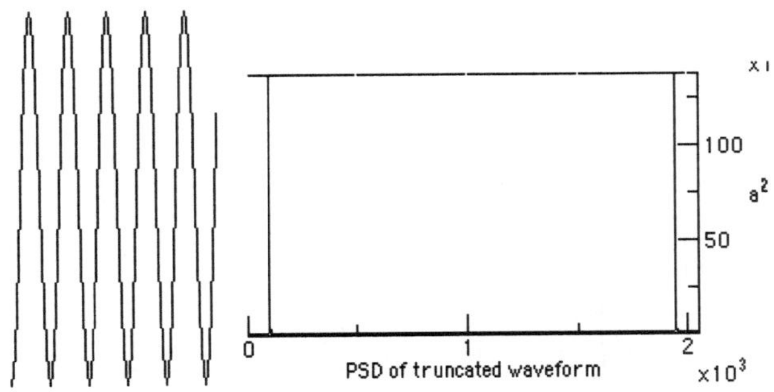

FIGURE 6.8 GRAPH OF THE SINE TONE AT 400 HZ WITH PSD.

Figure 6.9 shows the sine wave after noise is added.

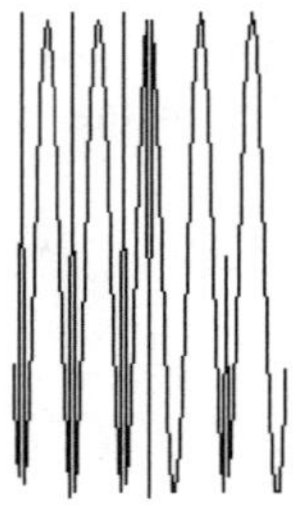

FIGURE 6.9 SINE WAVE AFTER THE ADDITION OF NOISE.

The PSD of the sine wave plus noise is shown in Figure 6.10.

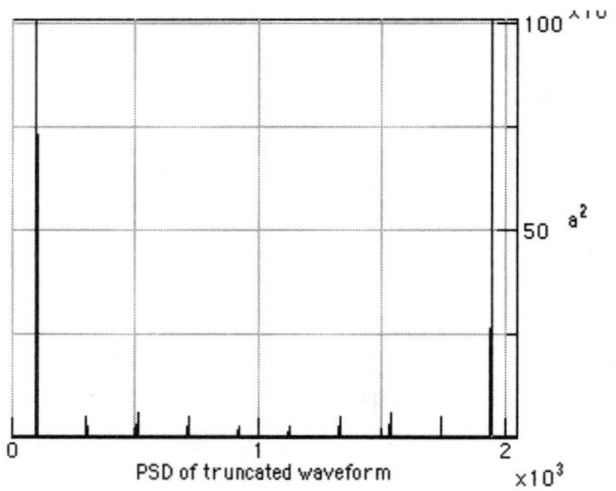

FIGURE 6.10 THE PSD OF THE SINE WAVE PLUS NOISE.

NOTE

Figure 6.10 shows the clean spectral break between the noise and the sine wave. The removal of noise from such a waveform is performed with a trivial frequency-based amplitude detector.

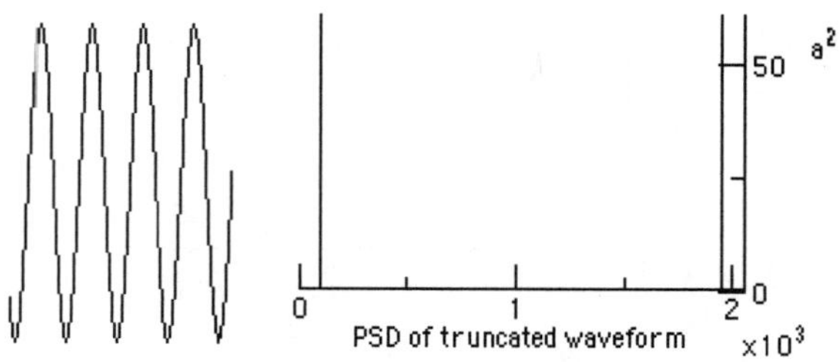

FIGURE 6.11 THE RECONSTRUCTED WAVEFORM AND ITS PSD.

Figure 6.11 should be the reconstructed sine wave with its PSD. Nose removal is not nearly as straight forward when waveforms have complex harmonic content.

Spectral Leakage of the DFT

Recall that in Equation 6.6 we can compute the frequency of the kth sample of a spectrum.

$$f_k = \frac{k}{N\Delta t}$$

For example, when the number of samples, N = 2,048, the smallest change in frequency is 8,000/2,048 = 3.9 Hz. For the spectrum shown in Figure 6.5, the central point is given by $N/2$ = 2,048/2 = 1,024. By Equation 6.6, therefore, we compute the highest frequency that can be represented by 2,048 samples (with 8KHz sampling rate) as (8,000/2,048) *1,024 = 4,000 Hz. Also, we can solve for any sample using Equation 6.25:

$$k = Nf_k / f_s \tag{6.25}$$

Thus, for the 400-Hz saw wave spectrum shown in Figure 6.8, we expect the maximum amplitude to occur at $k = Nf_k / f_s$ = 2,048*400/8,000 = 102.4. How can the FFT (or DFT, for that matter) represent energy at k = 102.4? The answer is that the energy is spread around k=102 and 103. Often the kth array element in the frequency domain is referred to as a frequency *bin*. Because the 400-Hz tone is in both bins 102 and 103, the phenomenon is called *spectral leakage*. The problem with spectral leakage is that it is different for different frequencies. For example, at 390.625 Hz, k = 2,048*390.625/8,000 = 10, *exactly*. Thus, there is no spectral leakage for some frequencies, and a great deal of spectral leakage for others. This is the frequency-domain rationale for spectral leakage.

The time-domain rationale for spectral leakage is that, for a waveform that is not periodic in time, the temporal effect of the sample window becomes visible in the Fourier transform [Walker].

A common procedure known as *windowing* predistorts the input samples so that the spectral leakage is evened out (spreading on-bin signals more and off-bin signals less). Windowing is typically performed with one of a variety of possible filters (Cesáro, Hann, Hamming, Parzen, Welch, and so on). It is beyond the scope of this book to cover all the filters. The popular

Hann filter is applied in the sample domain by a complex multiplication. When the input signal is real (as in audio), only a single multiplication is required. The equation for the Hann filter is as follows:

$$w_j = \frac{1}{2}\left[1-\cos\left(\frac{2\pi j}{N-1}\right)\right] \quad j \in [0 \ldots N-1] \tag{6.26}$$

Figure 6.12 shows a 256-sample version of the Hann window, in the frequency domain, generated by the DiffCAD program.

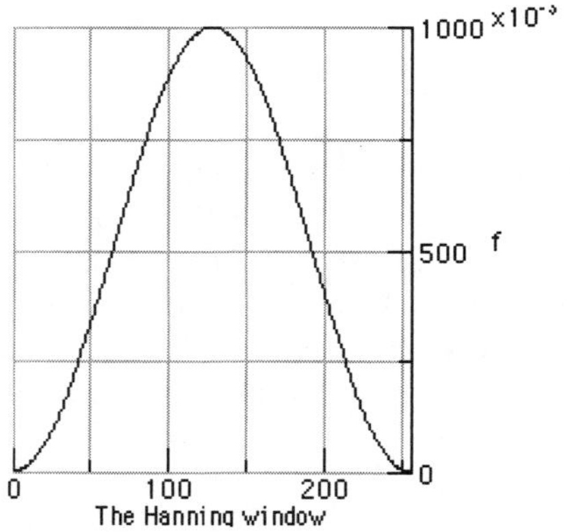

FIGURE 6.12 THE HANN WINDOW.

The main point of windowing is that it reduces the amplitude of the samples at the beginning and end of the window. The following code, from the *AudioFrame* class, will make and graph the Hann window:

```
public void makeHann() {
double window[];
     window = makeHann(256);
     Graph.graph(window,
                "The Hann window","f");
}
```

The method returns an array of doubles that can be stored for later use, permitting reuse of the window for production code.

```
public double[] makeHann(int n) {
     double window[] = new double[n];
```

```
     double arg = 2.0 * Math.PI/ (n - 1);

     for (int i = 0; i < n; i++)
          window[i] = 0.5 - 0.5 * Math.cos(arg*i);

       return window;

}
```

To apply the window to the samples, we have devised a *windowArray* method in the *AudioFrame* class:

```
public void multHann() {
     double[] r_d = getTruncatedDoubleData();
     double[] window = makeHann(r_d.length);

     windowAudio(r_d, window, "hann");
}

public void windowArray(double window[],double r_d[]) {
     for (int i = 0; i < window.length; i++) {
          r_d[i] *= window[i];
     }
}
```

The primary difference between the various windows is the way a window tapers off at the ends of the samples. The Bartlett window (Figure 6.13) is a linear window that is described by

$$w_j = \begin{cases} \dfrac{2j}{N-1} & j \in \left[0 \ldots \dfrac{N-1}{2}\right] \\ 2 - \dfrac{2j}{N-1} & j \in \left[\dfrac{N-1}{2} \ldots N-1\right] \end{cases} \qquad (6.27)$$

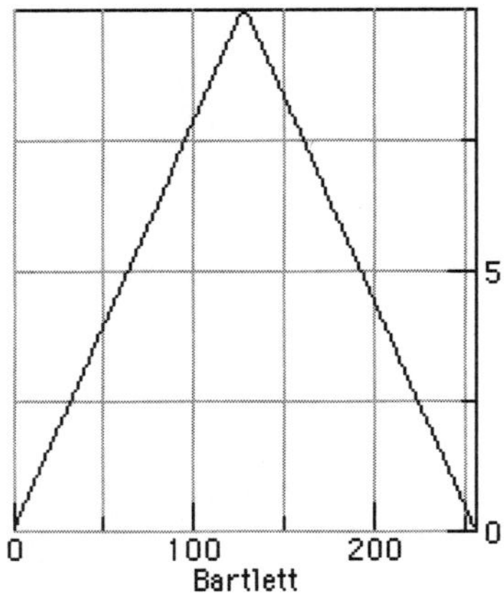

FIGURE 6.13 THE BARTLETT WINDOW.

The Bartlett window is generated by the *makeBartlett* method:

```
public double[] makeBartlett(int n) {
     double window[] = new double[n];
     double a = 2.0 /  (n - 1);
     for (int i = 0; i < n/2; i++)
          window[i] = i * a;
     for (int i = n/2; i < n; i++)
          window[i] = 2.0 - i * a;
       return window;
}
```

We propose a window that is zero at the sample endpoints and also has zero first and second derivatives at both the endpoints and the center. Such a window can be formulated with two fifth-order polynomials (called *quintics*). We present, for your interest, some Maple code for deriving a quintic-based window.

Maple is a symbolic manipulator that can help with some math problems. In Maple, we let *dy* and *ddy* be the first and second derivatives of the polynomial with respect to the dilation parameter *u*. The dilation parameter will be varied between zero and 1, inclusive. The amplitude of the quintic will vary from *y*0 to *y*1. Maple's optimization routine can output a procedure for computing quintics in minimum CPU time. What follows is a Maple procedure for generating the Lyon window. See [Lyon 91] for more information and an application of the quintic to maneuvering.

```
restart;y:=a5*u^5+a4*u^4+a3*u^3+a2*u^2+a1*u+a0:
> dy:=diff(y,u):
> ddy:=diff(dy,u):
> a0:=y0:
> a1:=solve(y1=subs(u=1,y),a1):
> a2:=solve(0=subs(u=0,dy),a2):
> a3:=solve(0=subs(u=1,dy),a3):
> a4:=solve(0=subs(u=0,ddy),a4):
> a5:=solve(0=subs(u=1,ddy),a5):
> readlib(C):
> C(y,optimized);
      t2 = u*u;
      t3 = t2*t2;
      t11 = (6.0*y1-6.0*y0)*t3*u+(-15.0*y1+15.0*y0)*t3+(10.0*y1-10.0*y0)*t2*u+y0;
```

The following code makes the Lyon window:

```
public double y5(double y0, double y1, double u) {
     double t2 = u*u;
     double t3 = t2*t2;
     return
          (6 * y1 - 6 * y0) * t3 * u +
          (-15 * y1 + 15 * y0) * t3 +
          (10 * y1 - 10 * y0) * t2 * u + y0;
```

The formula for the Lyon window follows:

$$w_j = \begin{cases} 6u^5 - 15u^4 + 10u^3 & u = \dfrac{2j}{N-1},\ j \in \left[0 \ldots \dfrac{N-1}{2}\right] \\ -6u^5 + 15u^4 - 10u^3 + 1 & u = \dfrac{2j+1-N}{N-1},\ j \in \left[\dfrac{N-1}{2} \ldots N-1\right] \end{cases} \tag{6.28}$$

The easier way to formulate (and compute) the dilation parameter, u, is by incrementing it by $2/N$:

```
public double[] makeLyon(int n) {
     double window[] = new double[n];


     double u = 0.0;
     double du = 2.0/n;
     for (int i = 0; i < n/2; i++) {
          window[i] = y5(0,1.0,u);
```

```
            u += du;
        }
        u=0;


        for (int i = n/2; i < n; i++) {
            window[i] = y5(1.0,0.0,u);
            u += du;
        }
          return window;


}
```

The Lyon and Hann windows are shown in Figure 6.14.

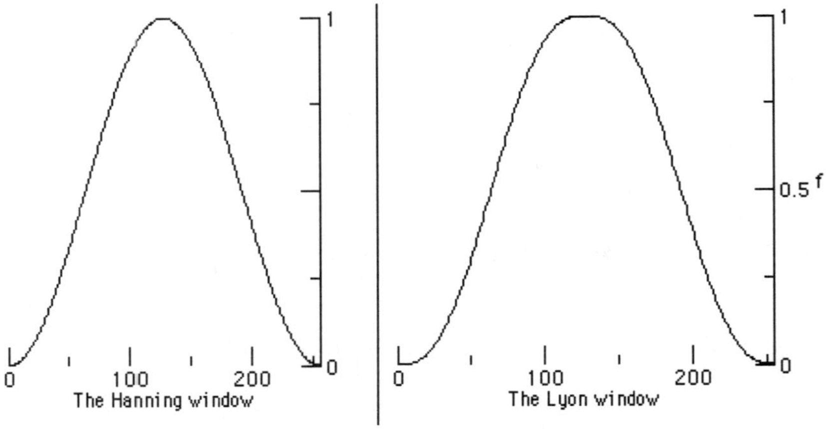

FIGURE 6.14 THE LYON AND HANN WINDOWS COMPARED.

NOTE

More study is needed to elaborate on the design trade-offs between the Lyon window and other popular windows. The most remarkable thing about it is that it has zero first and second derivatives at the center and endpoints. The Lyon window also has more taper at the side lobes than the Hann window and so may further reduce aliasing error.

The High-Pass Filter

One simple way to design a high-pass filter is to take an FFT on the input samples and then multiply the spectrum by the filter envelope. For example, we can zero out the low frequencies using a rectangular passband function, such as the one shown in Figure 6.15. Figure 6.16 shows a saw wave with its PSD. The PSD is multiplied by the characteristic function of the high-pass folter shown in Figure 6.15. The resulting PSD is shown on the right in Figure 6.17. On the left in Figure 6.17 is the reconstructed filtered wave form.

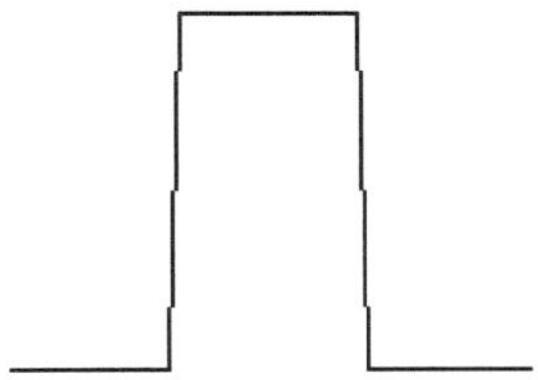

FIGURE 6.15 A HIGH-PASS FILTER (HIGH-FREQUENCIES ARE TOWARD THE CENTER).

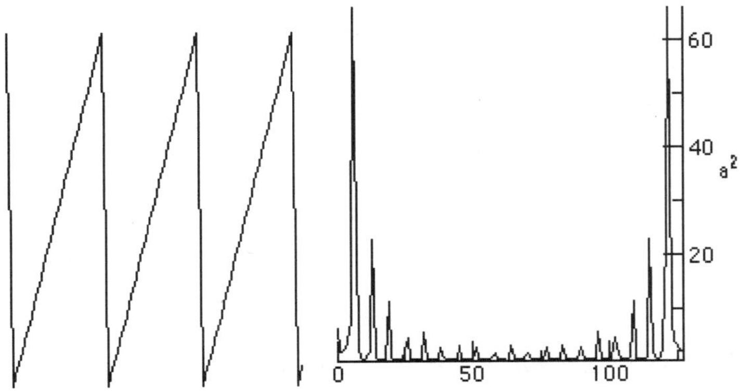

FIGURE 6.16 THE SAW WAVE AND ITS PSD.

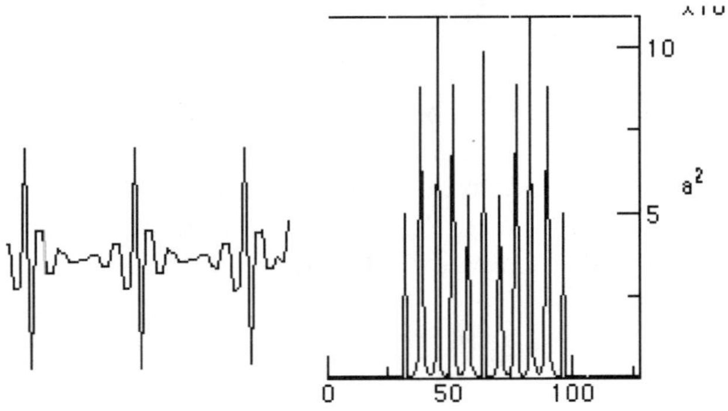

FIGURE 6.17 THE HIGH-PASS–FILTERED SAW WAVE AND ITS PSD.

NOTE

For the PSDs depicted in this chapter, the higher frequencies are toward the center of the graph.

For the determination of the performance of the various windows, we follow [Embree] and plot the windowed waveform alongside the log (in dB) of the PSD. We use the dB log and zoom into the first 200 samples of the PSD to make smaller details clear. We compute the graph using this code:

```
public void graphPSD() {
    double psd[] = fftInstance.computePSD();
    double shortPsd[] = new double[200];
    for (int i =0 ; i < shortPsd.length; i++) {
         shortPsd[i] = 10*Math.log(Math.sqrt(psd[i]));
    }
       Graph.graph(shortPsd, "spectral mag","a dB");
}
```

Figure 6.18 shows a sine wave with a rectangular window. The sine wave is 400 Hz with 8,000-Hz sampling. This is often expressed as a relative frequency of 400/8,000 = 0.05.

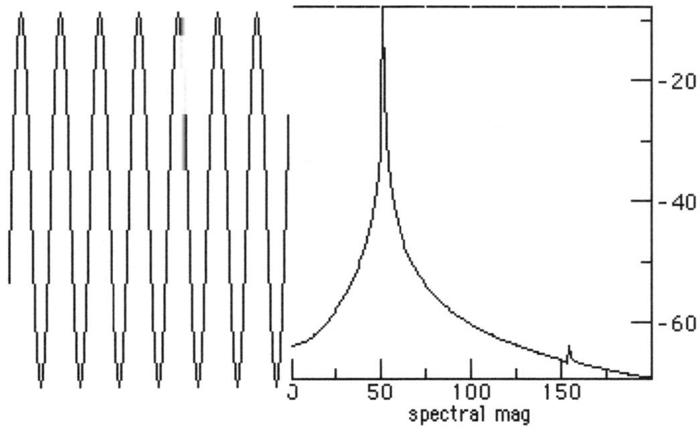

FIGURE 6.18 A RECTANGULAR WINDOW AND SPECTRAL DB LOG.

Figure 6.19 shows the Hann window with the spectral log magnitude.

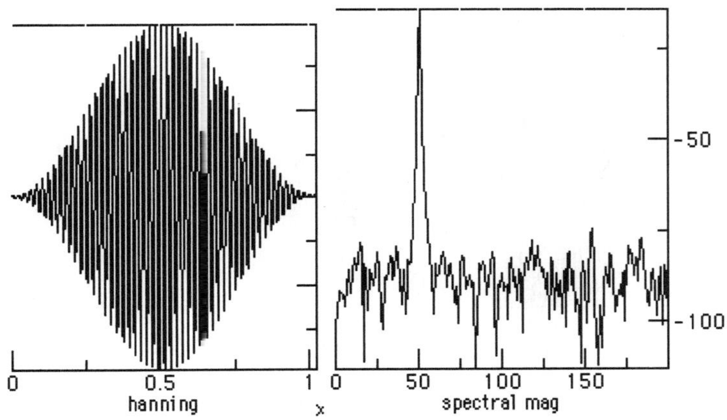

FIGURE 6.19 THE HANN WINDOWED DATA WITH THE FFT RESULT.

The triangular windowed input of the Bartlett window and the PSD are shown in Figure 6.20.

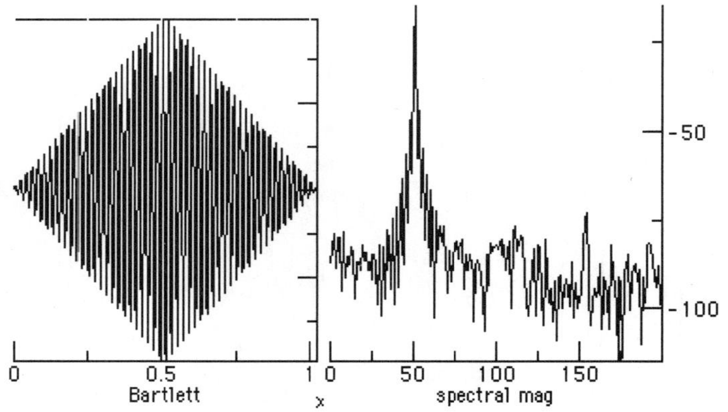

FIGURE 6.20 THE BARTLETT WINDOWED DATA AND THE PSD.

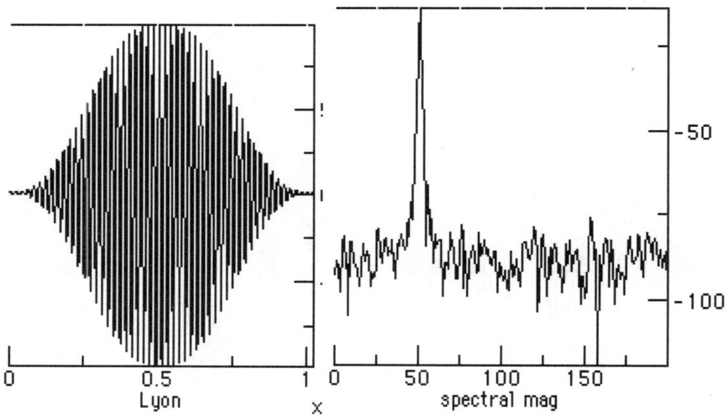

FIGURE 6.21 THE LYON WINDOW AND PSD.

Figures 6.20 and 6.21 show that the Lyon window has somewhat better side lobe performance (lower spectral leakage) than the Bartlett window. Also, the main lobe in the Lyon window is narrower than that of the Bartlett window (lower noise bandwidth).

Figure 6.22 shows the spectra for the Lyon and hann windows on the left and right, for comparison.

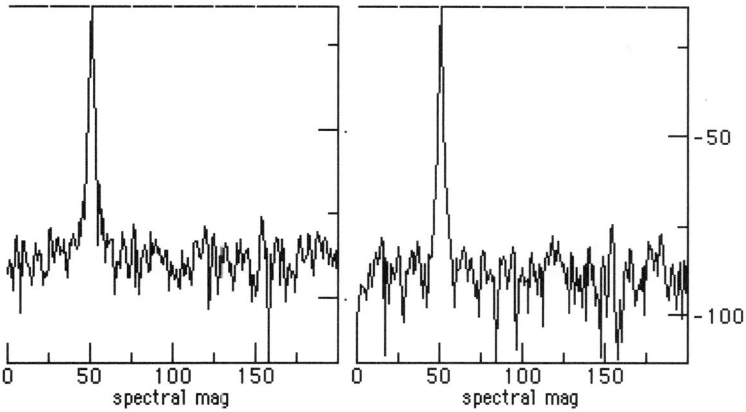

FIGURE 6.22 SPECTRA OF LYON (LEFT) AND HANN WINDOWS.

The Lyon and Hanning windows appear to be very close spectrally. The Lyon window appears to have a few dB better side lobe performance. The main lobe is slightly wider, however, and so has a slightly larger equivalent noise bandwidth. For more information on windowing, see [Mitra].

Frequency Shifting Using the FFT

To shift the pitch of a time-domain signal, we take the FFT, perform the high-pass filtering shown in the preceding section, shift the spectrum lower, and then perform the IFFT. Recall that the FFT produces a real and an imaging output. Also recall that *fftInstance* is a class variable in the *AudioFrame* class. Thus, we design our pitch-shifter as one of many possible spectral modifications that can be performed by the user before the IFFT is taken. The plan is to work on bins 0..*N*/2 first and then to copy the bins about the *N*/2 point in the spectrum, assuming that the left-hand and right-hand sides are symmetric (as is always the case for real signals). The code for the pitch-shift follows:

```
public void pitchShift() {
     fftInstance = new FFT();
     double[] r_d = getTruncatedDoubleData();
     int N = r_d.length;
     double[] i_d = new double[N];
     int N_on_4 = N/4;
```

```
        fftInstance.forwardFFT(r_d, i_d);
        // shift data down
        for (int i = 0; i < N_on_4; i++) {
             r_d[i] = r_d[i + N_on_4];
             i_d[i] = i_d[i + N_on_4];
        }
        for (int i= N_on_4; i < N/2; i++) {
             r_d[i] = 0;
             i_d[i] = 0;
        }


        // reflect about center, assuming a real signal
        int i,j;
        for (i=0,j=N-1; i < N/2; i++, j–) {
             r_d[j] = r_d[i];
             i_d[j] = i_d[i];
        }


        fftInstance.reverseFFT(r_d,i_d);


        ulc = new UlawCodec(r_d);


        ulc.play();


}
```

The result for synthetic tones, rich in harmonics, is to filter out some of the lower frequencies and to reduce the upper harmonic content.

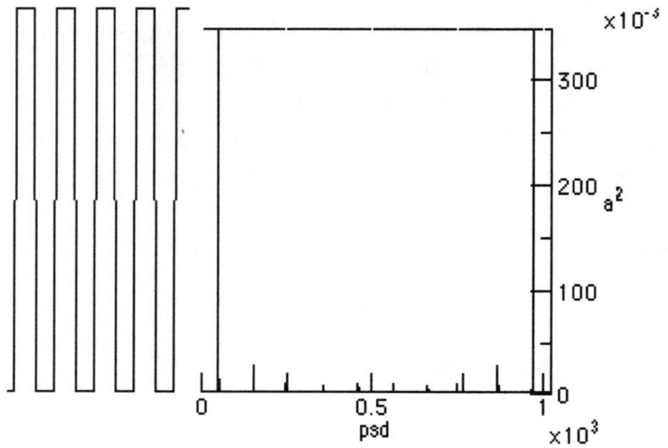

FIGURE 6.23 THE SQUARE WAVE AND ITS PSD.

Figure 6.23 shows a square wave and its PSD. Figure 6.24 shows the result of shifting the harmonics of the square wave away from the center of the PSD on the right. On the left of Figure 6.24, we4 see the reconstructed waveform in the time domain.

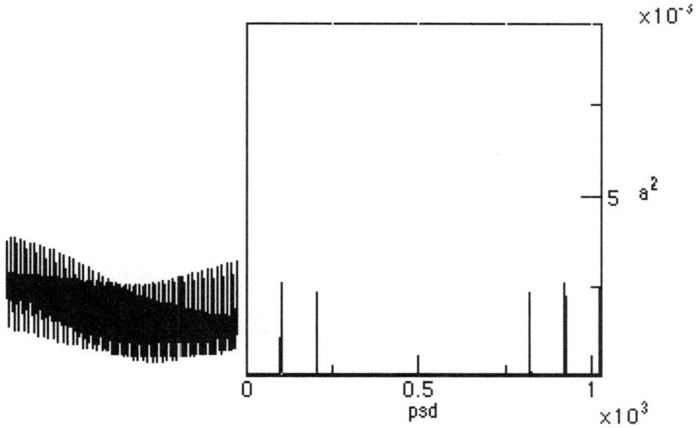

FIGURE 6.24 THE PITCH-SHIFTED SQUARE WAVE AND ITS PSD.

Resampling and the FFT

Resampling a one-dimensional waveform is a common way to perform time-compressed speech. One easy resampling technique is to perform Fairbanks sampling and throw away every other sample [Fairbanks].

The resampling method performs a 2:1 subsampling in the time domain. The code for the *resample* method follows:

```
public void resample() {
     double[] r_d = getTruncatedDoubleData();
     int N = r_d.length;
     double [] resampled = new double[N/2];
     for (int i=0; i < N/2; i++)
          resampled[i] = r_d[2*i];


     ulc = new UlawCodec(resampled);


     ulc.play();

}
```

Figure 6.25 shows a saw wave and its PSD. Compare Figure 6.25 with Figure 6.26 which shows the saw wave and PSD after the 2:1 subsampling algorithm has been applied.

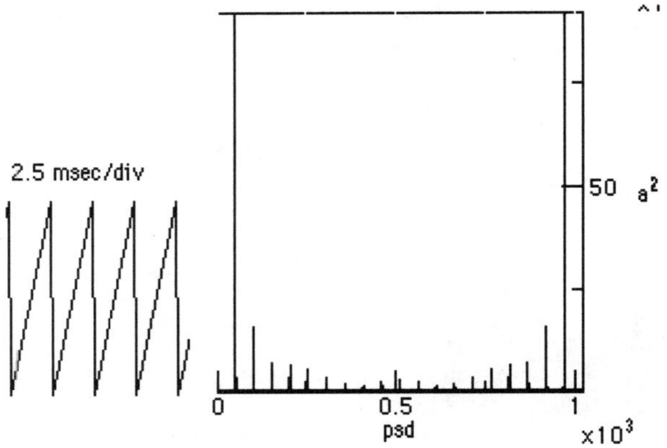

FIGURE 6.25 THE SAW WAVE AND PSD BEFORE SUBSAMPLING.

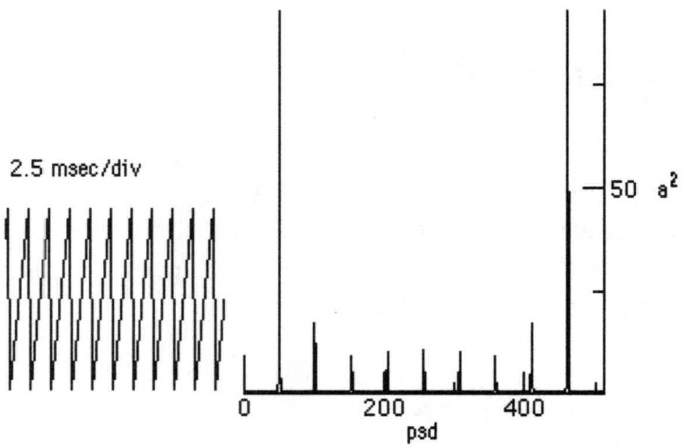

FIGURE 6.26 THE SAW WAVE AND PSD AFTER SUBSAMPLING.

The 2:1 subsampling has doubled the pitch of the harmonics and halved the number of available samples.

Centering the FFT

A great many books show that the lowest frequency is toward the center of the PSD when taking an FFT because the process of centering the FFT. The FFT is centered by replacing the sample data with a value that changes from positive to negative when the sample value goes from zero to 1. This can be described by the formula:

$$v_k = v_k(-1)^k \tag{6.29}$$

for the real sample on input. After the IFFT (or IDFT), the formula must be applied again [Myler].

The centered PSD for a pulse is shown in Figure 6.27.

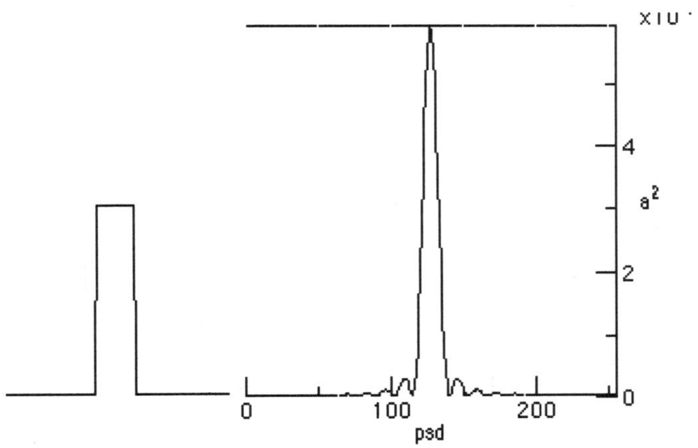

FIGURE 6.27 A PULSE WITH A CENTERED PSD.

You can modify the *forwardFFT* in the *AudioFrame* to perform this method:

```
public void forwardFFT(double in_r[], double in_i[])  {
     int id;
```

```
    int localN;
    double wtemp, Wjk_r, Wjk_i, Wj_r, Wj_i;
    double theta, tempr, tempi;
    int ti, tj;


    int numBits = log2(in_r.length);
    if (forward) {
         centering(in_r);
    }
```

The *centering* method computes Equation 6.28 on the real part of the sample data. Centering can be implemented as follows (in the *FFT* class):

```
private void centering(double r[] ) {
     int s = 1;
     for (int i = 0; i < r.length; i++) {
          s = -s;
          r[i] *= s;



     }
}
```

The *reverseFFT* implements *centering* after the IFFT has finished:

```
public void reverseFFT( double in_r[], double in_i[]) {
     forward = false;
     forwardFFT(in_r, in_i);
     forward = true;
     centering(in_r);
}
```

Without centering, the PSD is shown with the lowest frequencies on the edges (Figure 6.28). We have adopted this convention for the one-dimensional PSD display (except when explicitly marked otherwise). We are amazed that few books speak about the uncentered magnitude Fourier spectrum and how to correct it. Left uncentered, the FFT and DFT produce results that can match the outputs of other FFT and DFT implementations (as in [Moore]). As a result, we felt it best to leave the spectrum uncentered, at least for the one-dimensional FFT.

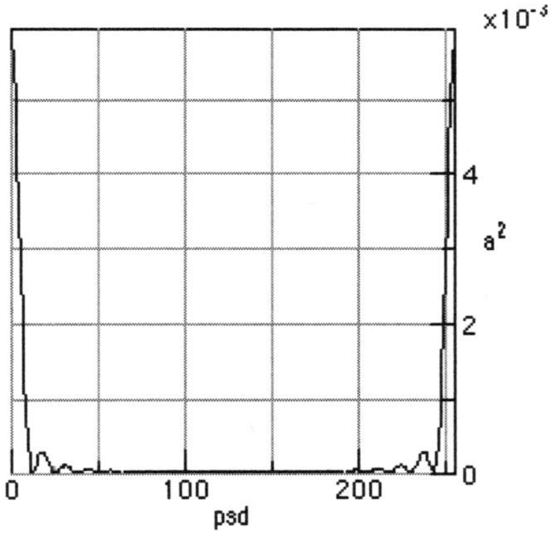

FIGURE 6.28 AN UNCENTERED PSD.

Summary

The close of this chapter marks a sad time for us. Just as things are getting fun, we move to the image processing section of the book (which we hope will be even more fun). Combining a pitch shifter with a resampler to compress speech is not a new idea. In fact, it can be used to help perform skimming on recorded speech (a topic of current research) [Arons].

The introduction to the DFT, IDFT, FFT, and IFFT is not new either, and the FFT is probably not the fastest. For the fastest Fourier transform in the west, see http://theory.lcs.mit.edu/~fftw. It may well be the fastest, but it may also rank as one of the most complex of implementations. It is still $O(N \log N)$, but it has a very low constant time. This link also has pointers to public domain software for performing FFTs (including a mixed radix FFT).

As far as we know, the derivation of the Lyon window, using Maple, is new. The Lyon window (two quintics with flat ends) is not new, although its application as a window for signals probably is. In the past, the quintic was viewed as a curvature controlled trajectory for maneuvering a car [Lyon 90]. We make no claim as to the suitability of the Lyon window for signal processing, because the window has not been thoroughly investigated. It poses a topic for future research.

Chapter 7

An Introduction to Image Processing

From her whole frame—an atmosphere which quite
Arrayed her in its beams, tremulous and soft and bright.
—*The Revolt of Islam*, A Poem in Twelve Cantos.

Canto Eleventh.
Peter Packed a Pickled Pixel
Its color was that of hash
for every program that displayed
This pixel it would crash
—*D. Lyon*

Image processing is a kind of digital signal processing that occurs on a two-dimensional array of sampled data. For example, it has been shown that image processing techniques can be used to compress DEM (digital elevation map) data [Franklin]. DEM data consists of a 2-D array of ranges, sometimes called a *range image*.

Often, the input image is acquired by a sensor that detects energy—typically, light energy. For the purpose of this chapter, we assume that we will sample an image using a uniform grid over which we perform the spatial sampling of the energy.

When we speak about the resolution of the image, we talk about the size of the array used to store the *significant* image data. For example, an array with 640 columns and 480 rows can contain double-precision floating-point numbers; if only 8 bits of information is significant, then the image resolution is 640×480 by 8 bits per pixel.

Image processing has applications in the areas of art, science, industry, government, and space [Holzmann] [Pratt]. Branches of image processing (such as digital image warping) have found applications in remote sensing, computer vision, special effects, and computer graphics [Wolberg].

Video Cameras and Scanners

Energy used to form an image typically starts as an analog signal (as in the one-dimensional DSP case). The primary difference is that the 2-D transducer detects the power in the incident energy and its relative direction.

Several techniques are commonly used to determine the relative direction of energy. A mechanical scanning technique (such as radar) physically moves a sensor to obtain a range image. Some flatbed scanners have a mechanical arm, called a *platen*, that contains a linear solid-state array. The platen is moved over an illuminated image. Another example of a mechanical scanner is a drum scanner. Typically a high-end scanner used in prepress applications, a drum scanner spins a drum on which art-work has been mounted.

Electronic scanning is performed by electron-beam deflection via an electric or magnetic field. The electron beam scans the photosensitive surface, which in turn alters the beam current in proportion to the incident light. Tube cameras are an example. The semiconductor industry has produced solid-state alternatives to tube cameras that dominate, both in cost and in performance specifications, all but niche markets. As a result, tube cameras have fallen out of favor; professional and consumer users typically favor solid-state cameras.

The primary solid-state camera in use today is the CCD (charged coupled device) camera, which was invented by Willard S. Boyle and George E. Smith at Bell Labs in the early 1970s. For semiconductor-based sensors, such as CCD cameras, a scanning array has individual sensing elements, each of which corresponds to a pixel. The size of the elements ranges from 7 to 14 μm. Recall that 1 μm (also called a micron) is 1×10^{-6} meters. The wavelength of a He-Ne laser (a red light) is 638 nm (nm = nanometers = 1×10^{-9} meters).

Figure 7.1 shows an image with an increasing gray level and a single white line that has been resampled from its center.

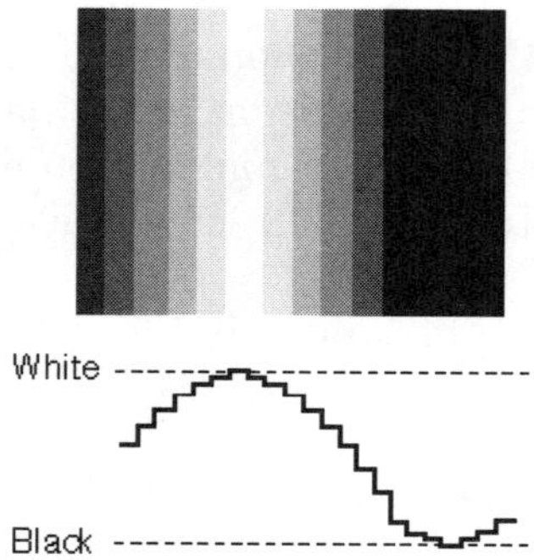

FIGURE 7.1 A STEPPED GRAY WEDGE AND THE OUTPUT OF A LINEAR SCANNING ARRAY.

Flatbed scanners typically have high-resolution linear CCD elements with a mechanically deflected platen that gives them higher resolution than hand-held cameras. For this reason, people often take a picture with a film camera and then scan the image to obtain a high-resolution scan. Film ranges in quality from common to laboratory grade. Common grade film (such as 35 mm Kodak 5369) has a grain size that yields 100 lines per mm (10,000 nm). Laboratory grade film yields about 1000 lines per mm (1,000 nm) [Kodak]. The high-end 35-mm digital slide scanners yield 2,048 × 3,072 pixels with 36 bits per pixel (see http://www.davidmyers.com.au/rfs2035.htm for an example). This is about 6.2 million pixels per image. Electronic still cameras have CCDs that have 850 × 984 pixel resolution (for example, http://www.kodak.com/daiHome/pdf/dc120.pdf). Video cameras (image sequence digitization cameras) typically operate with CCDs that have resolutions of 640 × 480 24-bit pixels (for example, http://www.kodak.com/daiHome/pdf/dvc300.pdf).

Although film is 100 to 1,000 times higher in resolution per unit of imaging area than CCD cameras, another factor is total size. The common size for film is 35-mm, but larger format negatives are available for special applications (70-mm movie film, 127-mm portrait cameras, large format X-ray film, panorama cameras, and so on). These large formats can greatly increase the number of pixels available. The other factor is cost; the price of digital still-image cameras appears to be dropping quickly. Also, it is not clear that people need the high resolution that film offers. We have found that students have no idea how much data can be produced by a high-resolution scan of a large image. For example, a 3 × 5 inch color photo scanned at 24-bits per pixel and at 1200 pixels per inch produces 3*5*1,200*1,200*3 = 64MB of data (MB = megabyte; Mb = megabit). We have found students turning up the virtual memory to 450MB when running Photoshop so that they can scan 8 × 10 photos at 1,200 pixels per inch (8*10*1,200*1,200*3 = 345MB). Some students have even said that they need such images for their home pages (no, really).

The Observer Interface

Although the Observer interface may seem out of place here, it is a central class for understanding how the *Image* class works. The *Observer* interface resides in the *java.util* package and is used to require implementation of methods that are needed to maintain consistency in an object-oriented environment. The relationship between the class that implements the *Observer* interface and class that extends the *Observable* class is the relationship between a view and a model (sometimes called the model-view relationship). An example of the model-view relationship is that between the a car's gas tank and its fuel gauge. If the amount of fuel in the tank changes, the fuel gauge readout changes. In Java, the simulation of a fuel tank notifies the fuel gauge readout so that the screen is consistent with the underlying simulation of the fuel tank.

There is a directed flow of information between an instance of a class that extends *Observable* and the instance of the class that implements *Observer*. The *Observable* instance keeps an instance of a vector that lists all the *Observer* instances that have registered an interest in the state of the *Observable* instance. Whenever the *Observable* instance changes, it broadcasts an update message to each of the *Observer* instances that have registered. This relationship is shown in Figure 7.2.

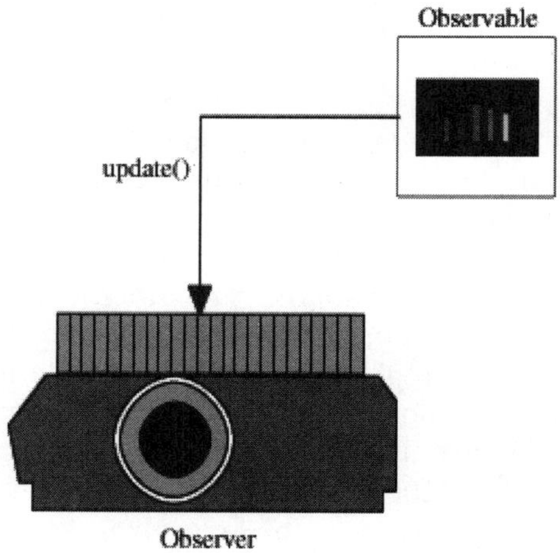

FIGURE 7.2 THE OBSERVABLE USES THE UPDATE METHOD TO TRANSMIT DATA TO THE OBSERVER.

A class that implements the *Observer* interface supports an *update* method that takes an *Object*-typed argument.

Interface Summary

```
package java.util;

public interface Observer {
    void update(Observable o, Object arg);
}
```

The Observable Class

The *Observable* class resides in the *java.util* package. An instance of the *Observable* class uses an instance of a list of the *Observer* instances that have registered themselves. The list is called an *ObserverList* and resides invisibly within the *java.util* package.

Class Summary

```
package java.util;
public class Observable {
    public synchronized void addObserver(Observer o)
    public synchronized void deleteObserver(Observer o)
    public void notifyObservers()
    public synchronized void notifyObservers(Object arg)
    public synchronized void deleteObservers()
    public synchronized boolean hasChanged()
    public synchronized int countObservers()
}
```

The NamedObservable Class

The *NamedObservable* class provides a mechanism to associate a *String* instance with the *Observable* instance.

```
package observers;
import java.util.*;
// The NamedObservable is just like an Observable
// except that it has a name property associated with
// every Observable instance.
public abstract class NamedObservable extends Observable {

private String name;
public synchronized void setName(String nm) {
     name = nm;
}

public synchronized String getName() {
     return name;
}
}
```

NOTE The *get* and *set* methods in the *NamedObservable* class are synchronized to prevent contention problems during multithreaded operation. It is common to see class variables, such as the *name* string, as being accessed only through synchronized methods. This is the reason for declaring the *name* string private.

The ObservableDouble and Observer-Observable Example

The DiffCAD program depends on the model-view paradigm and so has a package called *observers*. One of the classes in the *observers* package is *ObservableDouble*.

When the *setValue* method is invoked on an instance of the *ObservableDouble* class, the *setChanged* method is invoked and the *notifyObservers* method causes an *update* method to be broadcast to all the interested *Observer* instances.

```
package observers;
import java.util.*;

public class ObservableDouble extends NamedObservable {

// The value of interest
private double value;

public ObservableDouble(double newValue, String nm) {
     value = newValue;
     setName(nm);
}

public synchronized void setValue(double newValue) {
    if (newValue != value) {
       value = newValue;
       super.setChanged();
       super.notifyObservers();
    }
}

public synchronized double getValue() {
         return value;
   }
}
```

DoubleDialog and IntDialog

In the *observers* package are the *DoubleDialog* and *IntDialog* classes. When an instance of *IntDialog* is made, a display is generated like the one shown in Figure 7.3.

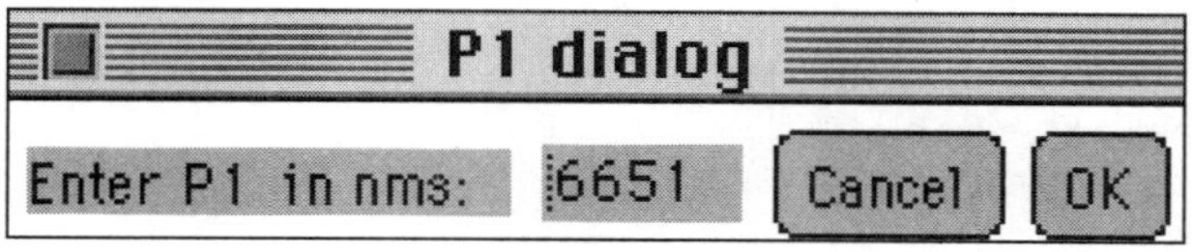

FIGURE 7.3 INTDIALOG WITH AN EMBEDDED OBSERVABLE ON DISPLAY.

The value of the *int* is initially used for the display in the text field. When the *int* is altered by the user, the *setValue* method causes the *observers* to be updated, dynamically altering the text field in the dialog box. The *IntDialog* class follows. (The *DoubleDialog* is almost the same except that it contains an *ObservableDouble* rather than an *ObservableInt*.)

```
package observers;
import java.awt.*;
import java.applet.*;
import java.io.*;

import gui.*;

public class IntDialog extends ClosableFrame {

    Label label1;
    IntTextField textfield;

    Button okButton;
    Button cancelButton;

    ObservableInt i;
    private static int offset = 25;

    public IntDialog(
        String dialog_title,
        String label,
        ObservableInt sample_int) {
     super(dialog_title);
     init( dialog_title,  label);
     set(sample_int);
     setForeground(Color.white);
```

```
    setBackground(Color.white);

   }

   void set(ObservableInt sample_int) {
        i = sample_int;

        textfield = new IntTextField(i);
        add("Right",textfield);
        textfield.reshape(89, 25, 64, 16);
        cancelButton = new Button("Cancel");
        add(cancelButton);
        cancelButton.reshape(13, 175, 88, 28);

        okButton = new Button("OK");
        add(okButton);
        okButton.reshape(130, 175, 63, 29);

        pack();
        show();
   }
  public void init(String dialog_title, String label) {
   setLayout(new FlowLayout());
   resize(250, 250);
   reshape(offset, offset,
            250, 250);
   offset = offset + 15;
   setResizable(false);

   //Initialize components
   label1 = new Label(label);
   add("Left", label1);
   label1.reshape(5, 19, 86, 23);
  }

void ok() {
        System.out.println("OK!");
        String s = textfield.getText();

        Integer i_Integer = new Integer(i.getValue());
             // default values
        try {
                i_Integer = Integer.valueOf(s);
                i.setValue(i_Integer.intValue());
            }
         catch (NumberFormatException e)  {
               System.out.println(
                  "IntDialog:ER! Not a number, using defaults");
```

```
                    // pick a reasonable default
                }
          return;
    }

 private void cancel() {
       System.out.println("Cancel!");
 }

 public boolean handleEvent(Event event) {
      if (Evt.match(event, 'o', okButton))
           ok();

      else if (Evt.match(event, 'k', cancelButton))
           cancel();
      return super.handleEvent(event);
 }
}
```

NOTE We realize that the user may want to make the *cancel* method perform a task other than print "Cancel." This would be a good spot to send an application-specific message or restore a value.

The following example of the use of *IntDialog* can be found in the *AudioFrame* class.

```
IntDialog spDialog = new IntDialog(
                 "start Position Of Samples To Graph dialog",
                 "Enter start position:",
                 startPositionOfSamplesToGraph
                 );
```

Where

```
     private ObservableInt startPositionOfSamplesToGraph;
```

is declared in *AudioFrame* as a class variable.

Dialog Boxes in the ImageFrame

The *ImageFrame* class, which resides in the *lyon.ipl* package, is used to hold and display instances that contain image data. It is also used to dispatch events and manage dialog boxes that prompt the user for parameters.

All the dialog boxes contain instances of *Observable* classes. Following is a list of the *Observable* class variables in the *ImageFrame* and the *Dialog* instance creation that depends on them. Lines 1 and 2 contain two *ObservableDouble* instances, each provided with a label (because they are named) and a value.

```
private ObservableDouble
        multKonst  = new ObservableDouble(1.0,"mult:");
private ObservableDouble
        addKonst  = new ObservableDouble(0.0,"add:");
private ObservableInt threshholdKonst
        = new ObservableInt(128, "Threshold:");
private ObservableInt scaleKonst
        = new ObservableInt(1, "Scale:");
private DoubleDialog multDialog
        = new DoubleDialog("Multiply Dialog",
          "* k", multKonst);
private DoubleDialog addDialog
        = new DoubleDialog("Add Dialog",
          "+ offset", addKonst);
private IntDialog threshDialog
        = new IntDialog("Threshold Dialog",
          "Threshold constant", threshholdKonst);
private IntDialog scaleKonstDialog
        = new IntDialog("Scale Dialog",
          "Integer Scale constant", scaleKonst);
```

When an instance of the *ImageFrame* class is made, the *IntDialog* and *DoubleDialog* instances appear on the screen. At that point, the user can override the defaults in the dialog boxes and click **OK**. The image processing operation is invoked and uses the updated values. For example:

```
public void linearComb() {
     pp.linearComb(multKonst.getValue(),addKonst.getValue());
     updateDisplay(pp);
}
```

The Image Class

In Java, an image is stored in an instance of the *Image* class, which resides in the *java.awt* package. The *Image* class is designed with the intention that images are produced by I/O bound systems (network-based systems). The *Image* class has provisions to take *ImageObserver* instances in many of its methods. The *ImageObserver* instance is registered for notification after an image becomes available.

Class Summary

```
package java.awt;
import java.awt.image.ImageProducer;
import java.awt.image.ImageObserver;

 public abstract class Image {
  public abstract int getWidth(ImageObserver observer);
  public abstract int getHeight(ImageObserver observer);
  public abstract ImageProducer getSource();
  public abstract Graphics getGraphics();
  public abstract Object getProperty(String name, ImageObserver observer);
  public static final Object UndefinedProperty = new Object();
 public abstract void flush();
}
```

Class Usage

The *Image* class uses the observer-observable model. The Image class is an abstract class and so can not be instanced. Recall from Chapter 3 that every *Component* implements an *ImageObserver*. This means that we can pass an instance of the *Component* to the *image.getWidth* invocation to obtain a callback. The callback will occur in the form of an *imageUpdate* method invocation on the interested *ImageObserver* instance.

Suppose the following variables are predefined:

```
Image image;
ImageObserver imageObserver;
ImageProducer imageProducer;
int height, width;
Graphics graphics;
String name;
```

To get the width of the image, use this code:

```
width = image.getWidth(imageObserver);
```

It returns -1 if the image is not ready and notifies *Observer* when the image becomes ready.

To get the height of the image, use this code:

```
height = image.getHeight(imageObserver);
```

It returns -1 if the image is not ready and notifies *Observer* when the image becomes ready.

To get the instance that produces the image pixels:

```
imageProducer = image.getSource();
```

To get a graphics instance for an off-screen image:

```
graphics = image.getGraphics();
```

To get an image property by name:

```
object = image.getProperty(name, imageObserver);
```

If the image is not available, *getProperty* returns null and notifies *imageObserver* later. *Image.UndefinedProperty* is a static final *Object* instance that is returned when the property is undefined.

To destroy all the resources used by an image that will probably not be used for a while, use this:

```
image.flush();
```

The ImageObserver Interface

The *ImageObserver* interface resides in the *java.awt.image* package. An instance of a class that implements the *ImageObserver* interface registers as having an interest in the contents of an *ImageProducer* instance. When the contents are altered, the *ImageObserver* instance is notified by the *ImageProducer* instance using the *imageUpdate* method invocation.

Class Summary

```
package java.awt.image;
import java.awt.Image;
public interface ImageObserver {
    public boolean imageUpdate(Image img, int infoflags, int x, int y, int width,
    int height);
```

```
    public static final int WIDTH = 1;
    public static final int HEIGHT = 2;
    public static final int PROPERTIES = 4;
    public static final int SOMEBITS = 8;
    public static final int FRAMEBITS = 16;
    public static final int ALLBITS = 32;
    public static final int ERROR = 64;
    public static final int ABORT = 128;
}
```

Image Instancing in the ImageFrame

The *ImageFrame* class has a *getImage* method that prompts the user for an input file and then opens the image file, waiting until it is read and the resources needed to store the *Image* instance are allocated. We have decided to defeat this design pattern to simplify the job of reading an image. We find the code to be harder to understand, write, and teach and prefer a simpler API for reading an image. To defeat the callback routine, we make a temporary *ImageProducer* that is invoked with *getImage*. Methods such as *getImage* greatly simplify user code, but they assume that the image resides on a low latency storage device. The *fileName* and *image* variables are class variables in the *ImageFrame* class. The *getImage* method follows:

```
public void getImage() {
     // Open the image by file name and wait for it!
     fileName = Futil.getReadFileName();
     System.out.println("Get image:"+fileName);
     try {
               image = getToolkit().getImage(fileName);
               waitForImage(this,image);
               }
          catch (Exception e) {
               System.out.println("Get Image could not open "+e);
          }
          // move the image, change its dimensions
          reshape(100, 100,
               image.getWidth(this), image.getHeight(this));


          //set image title
          setTitle(fileName);
          show();
}
```

The *waitForImage* method avoids the callback method and stops the thread before the *getImage* method returns. The *paint* method for the *ImageFrame* invokes the *drawImage* method on the *Graphics* instance, *g*.

```
public void paint(Graphics g) {
      g.drawImage(image,0,0,this);
}
```

The PixelPlane Class

The *PixelPlane* class resides in the *lyon.ipl* package and is used to give the user a high-level interface to pixel-based operations. This class permits you to process an array of data without having to follow the *ImageProducer-ImageConsumer* design pattern. This approach makes the software much easier to teach, write, and understand. We have found that Java programmers prefer the simpler array model for manipulating images. An instance of a *PixelPlane* class stores its pixels in a 1-D array of int. A *PixelPlane* instance does not do image processing, but is a handy way to access pixels and allocate *Image* instance memory.

The basic idea is that we do not need to keep an instance of an *Image* around. All the image processing occurs on and between instances of the *PixelPlane* class. If we need an *Image* instance (for display), we can always make one from an instance of the *PixelPlane* class. This approach—making *Image* instances on-demand—represents a typical trade-off between space and time. An *Image* instance takes up so much space that dynamic allocation can free enough space during processing to make otherwise unfeasible operations feasible (from a memory usage point of view).

Because Java does not have (as of this writing) any notion of row-major or column-major order, we cannot use doubly nested *for* loops to access 2-D arrays. To do so might cause disk thrashing. (This is particularly slow and is typical when memory is accessed nonsequentially.) To ease the burden of thrashing, a single-dimensioned array is stored internally in the *PixelPlane* instance. The 2-D array is implemented by multiplying the row index by the column index. Methods are also provided that permit linear indexing into the array, resulting in slightly higher speed for some image processing operations.

Class Summary

```
package lyon.ipl;
import futils.*;
import futils.utils.*;
import java.awt.*;
```

```
  import java.awt.image.*;
  import java.applet.Applet;
  public class PixelPlane {
  public int pels[]
  public PixelPlane(int w, int h)
  public PixelPlane(double _x, double _y)
  public PixelPlane copy()
  public Image makeImage()
  public int getLength()
  public int getHeight()
  public int getWidth()
  public boolean inrange(int x, int y)
  public int getRed(int i)
  public int getGreen(int i)
  public int getBlue(int i)
  public int getRed(int x, int y)
  public int getGreen(int x, int y)
  public int getBlue(int x, int y)
  public int getAlpha(int x, int y)
  public int MakePixel(int r, int g, int b, int a)
  public void setPixel(int x, int y, int r, int g, int b, int a)
  public void setPixel(int x, int y, int pel)
  public int getPixel(int x, int y)
  public int getPixel(int i)
  public void printSize()
}
```

Class Usage

Suppose the following variables are predefined:

```
PixelPlane pp, ppCopy;
int width, height;
double w, h;
int pixelArray[];
Image image;
boolean aBoolean;
int x,y;
int i, c, a, r, g, b, pel;
```

To make an instance of a *PixelPlane* that is *width* by *height*, use this code:

```
pp = new PixelPlane(width, height);
```

The *PixelPlane* constructor is overloaded to be used with double-type dimensions:

```
pp = new PixelPlane(w, h);
```

To get at the internal pixel array (this is not suggested but is permitted to maintain flexibility), use this code:


```
pixelArray = pp.pels;
```

To make a copy of the *PixelPlane* (this uses the *System.arraycopy* method, so it should be fast):

```
ppCopy = pp.copy();
```

To make an *Image* instance from a file (makes a standard file open dialog box), use this:

```
image = PixelPlane.openImage();
```

openImage is a static method, so you don't need to make a *PixelPlane* instance. To get the total number of pixels in the *PixelPlane* instance, use this:

```
int numberOfPixels = pp.getLength();
```

To get the *width* and *height*, in pixels, from the *PixelPlane* instance:

```
height = pp.getHeight();
width = pp.getWidth();
```

To see whether two (x,y) coordinates are in range:

```
aBoolean = inrange(x,y);
```

To get a color stored in an *int* from a location in the *PixelPlane* instance (using the linear array):

```
c = pp.getRed(i);
```

The color value, c, always varies from 0 to 255.

```
c = pp.getGreen(i);
c = pp.getBlue(i);
```

To get the same color values using the (x,y) coordinates:

```
c = pp.getRed(x, y);
```

Each access into the *PixelPlane* costs one multiplication. After Sun releases an API that stores multidimensional arrays, in known order, the multiplication can be eliminated, speeding pixel access:

```
c = pp.getGreen(x, y);
c = pp.getBlue(x, y);
c = pp.getAlpha(x, y);
```

A pixel is stored in a packed-in, 4 bytes per pixel, ARGB (alpha, red, green, blue) format. To pack a pixel:

```
pel = pp.MakePixel(r, g, b, a);
```

To set a pixel at location (x,y):

```
pp.setPixel( x, y, r, g, b, a);
```

To set a pixel equal to the packed *pel* at location (x,y):

```
pp.setPixel(x, y, pel);
```

To get a packed pixel at location (x,y):

```
pel = pp.getPixel(x, y);
```

To get a packed pixel from the linear array at location *i*:

```
pel = pp.getPixel(i);
```

To print the size of the *PixelPlane* instance to the console:

```
pp.printSize();
}
```

The ProcessPlane Class

The *ProcessPlane* class resides in the *lyon.ipl* package. It consists of an extension to the *PixelPlane* class. By creating instances of the *ProcessPlane* class, we add methods to the *PixelPlane* instance that provide for elementary image processing services. The services themselves are not profound, but the way they are implemented in Java is of interest.

Class Summary

```
package lyon.ipl;
import java.awt.*;
public class ProcessPlane extends PixelPlane {
   public ProcessPlane (double x_, double y_)
   public ProcessPlane (int x_, int y_)
   public void cornergray() {
   public ProcessPlane scale(int scale) {
   public void threshold(int konst)
   public void diagGray()
   public ProcessPlane edge()
   public void linearComb(double konstD, double akD)
   public void Subimage(ProcessPlane pp)
   public void makeGray()
   public void randResample()
   public void shadow()
   public void negate()
}
```

Class Usage

Suppose the following variables are predefined:

```
ProcessPlane pp, pp2;
int height, width;
double heightD, widthD, a, b;
int scale, konst;
```

To make an instance of a *ProcessPlane* that allocates internal pixel storage for a *height* × *width* by 32-color, RGBA (Red, green, blue, and alpha) image, use this code:

```
pp = new ProcessPlane(height, width);
```

To make an instance of a *ProcessPlane* that uses doubles for size and casts the doubles to *int* before allocating storage:

```
pp = new ProcessPlane(heightD, widthD);
```

To make an image of the previous image's size and fill it with a *cornergray* test pattern, as shown in Figure 7.4:

```
pp.cornergray();
```

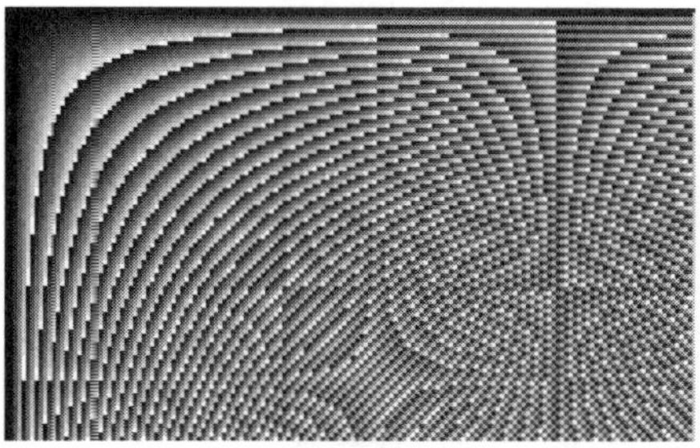

FIGURE 7.4 THE CORNERGRAY METHOD OUTPUT.

To scale an image by a positive integer using pixel replication:

```
pp.scale(konst);
```

To threshold an image using an integer value that ranges from 0 to 255, as shown in Figure 7.5:

```
pp.threshold(konst);
```

FIGURE 7.5 EFFECT OF THRESHOLD ON AN IMAGE.

To make a diagonal test pattern using the existing image's dimensions, as shown in Figure 7.6:

```
pp.diagGray();
```

FIGURE 7.6 THE EFFECT OF THE DIAGGRAY METHOD.

To make a single pixel-width wide edge from the leftmost edge of the image, at the center of a thick bright stripe, as shown in Figure 7.7, use this code:

```
pp.edge();
```

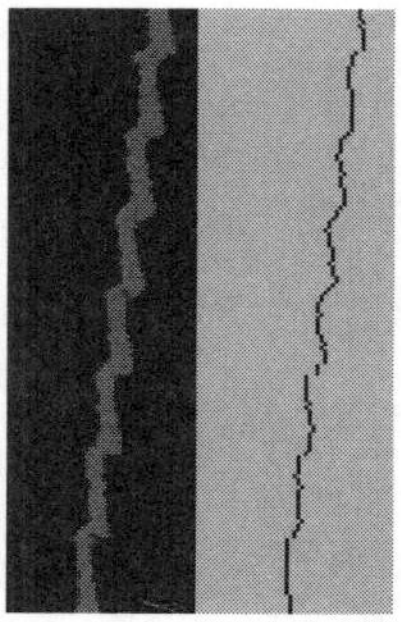

FIGURE 7.7 EFFECT OF THE EDGE METHOD.

The *edge* method uses a domain-specific edge detector that is useful for range-finding via diffraction. See the following section for more information about the implementation.

NOTE

This edge detection method will produce unsatisfactory results when applied to images outside the intended domain.

To perform the operation

$$p_i = ap_i + b \qquad \forall_i \in \left[i = 0 \ldots numberOfPixels\right]$$

where

$$p_i = \text{ ith image pixel}$$

```
pp.linearComb(a, b);
```

To subtract *pp2* from *pp*, leaving the result in *pp* (i.e., *pp = pp - pp2*):

```
pp.Subimage(pp2);
```

To make *pp* a gray image by copying the red plane to the green and blue planes:

```
pp.makeGray();
```

To perform a stochastic (random) resampling of the image, an effect illustrated in Figure 7.8:

```
pp.randResample();
```

FIGURE 7.8 THE EFFECT OF RANDRESAMPLE.

To perform a shadow mask on the image (as shown in Figure 7.9):

```
pp.shadow();
```

FIGURE 7.9 EFFECT OF THE SHADOW METHOD.

To negate the image (as shown in Figure 7.10):

```
pp.negate();
```

FIGURE 7.10 EFFECT OF THE NEGATE METHOD.

This section showed a collection of image processing services performed by instances of the *ProcessPlane* class. Although all the images are shown as gray-scale images, this is a limitation of the printing process. All the operations are 24-bit color operations.

Class Implementation: The negate Method

None of the image processing operations shown in this section is remarkable. Perhaps the more interesting issue is how the operations are implemented. We start with the simplest of the image processing operations, the *negate* method:

```
public void negate() {
     int r,g,b,a;
     for (int y=0; y < getHeight(); y++)
          for (int x =0 ; x < getWidth(); x++) {
               r = getRed  (x,y);
               g = getGreen(x,y);
               b = getBlue (x,y);
               a = getAlpha(x,y);
               setPixel(x, y, 255 - r,255 - g,255 - b, a);
          }
}
```

The *negate* method is added to the *ProcessPlane* class to enable any instance of a *ProcessPlane* to negate itself.

NOTE

Because we're extending the *PixelPlane* class, we inherit all its methods. Also, we do not negate the alpha plane, because that would make the image black.

In the *ImageFrame* class is a *MenuItem* instance called *negate_mi*:

```
MenuItem negate_mi     = addItem("[n]negate");
```

The *addItem* method adds the *MenuItem* instance to the main menu bar:

```
public MenuItem addItem(String itemName) {
     MenuItem mi = new MenuItem(itemName);
     m.add(mi);
     return(mi);
}
```

Finally, we handle the event using the *Evt*'s *match* method to select either the keyboard event or the menu selection event:


```
public boolean handleEvent(Event e) {
  if (Evt.match(e,negate_mi)) {
           negate();
          return true;
     }...
```

Class Implementation: The shadow Method

The *shadow* method is a filter that moves a 2 × 2 pixel window, one pixel at a time, across an image. It replaces the pixel at the upper-left corner of the window with the value of the pixel minus one-half the value of the pixel at the lower-right corner.

```
public void shadow() {
     int r,g,b,a;
     for (int y=0; y < getHeight()-2; y++)
          for (int x =0 ; x < getWidth()-2; x++) {
                r = getRed  (x,y);
                g = getGreen(x,y);
                b = getBlue (x,y);
                a = getAlpha(x,y);
                r = r + (127 - getRed  (x+2,y+2));
                g = g + (127 - getGreen(x+2,y+2));
                b = b + (127 - getBlue (x+2,y+2));
                setPixel(x,y,r,g,b,a);
          }
}
```

Class Implementation: The edge Method

The *edge* method is probably one of the most complex image processing methods in the *ProcessPlane* class. The basic idea behind the *edge* method is to run through the following steps:

1. Compute the average intensity, *a*, for pixel row, *y*.
2. Find the pixel, *xr*, that exceeds *a*.
3. Find the very next pixel, *xl*, that does not exceed *a*.
4. Compute the average of *xr* and *xl* and call it *xa*.
5. Store *xa,y* as an edge point.
6. Increment *y*.

7. If *y* > *imageHeight* then stop, else goto 1.

This type of ad hoc algorithm will always find the leftmost bright edge for a well-defined stripe. At the same time, it finds the centroid of the stripe. As the stripe gets closer to the left of the image, the program runs faster (because it has to search fewer pixels to find the stripe). Thus, good lighting and camera positioning are needed to make this algorithm run fast. The Java implementation of the *edge* method follows:

```
public ProcessPlane edge() {
     ProcessPlane pp = new ProcessPlane(getWidth(), getHeight());

     int r,a;
     int average; // average intensity
     for (int y = 0; y < getHeight(); y++)
     new_line: {
               average = 0;
               for (int x =0 ; x < getWidth(); x++)
                    average = average + getRed  (x,y);
               average = average / getWidth();
               for (int x =0 ; x < getWidth(); x++) {
                 r = getRed  (x,y);
                 a = getAlpha(x,y);
                 if (r > average) {
                      int start=x;
                      for (int i=x;i < getWidth(); i++) {
                           r = getRed  (i,y);
                           if (r < average) {
                                pp.setPixel(
                                     start + (i - start) / 2,
                                     y,255,255,255,a);
                                break new_line;
                                }
                           }
                      }
                 }
          }
     return pp;
}
```

Chapter 8

Digital Images and Image Formats

"This is the Wild West of the Information Age"

—*Bart Kosko*

The Databahn

As increasing numbers of users crowd onto the Internet information thoroughfare, it takes longer to get data from point A to point B. As never before, we need to shrink information content before it is exchanged or transferred. An uncompressed digital bitmap image of 640x480 pixels with 256 colors takes up 307Kb, or nearly one-third of a megabyte. It can be frustrating to wait for such an image file to load into your browser from a Web site. The enormous size of digital images has motivated the creation of a number of space -saving image and file formats.

In this chapter, we discuss the how and why of image formats—there are many of them—as well as compression techniques. You will learn about the latest and greatest formats for use on the Internet, including details of the formats that are supported in DiffCAD.

A Bird's–Eye View of Image Formats

There are three broad categories of image formats: bitmap formats, vector formats, and other formats.

Bitmap Formats

In describing an image, you can resort to several levels of detail. At the lowest level of abstraction, you can describe every element (pixel) of the image. Such an image is referred to as *bitmapped*, because it is a map of bits (or pixels). A bitmap image and its associated data are shown in Figure 8.1.

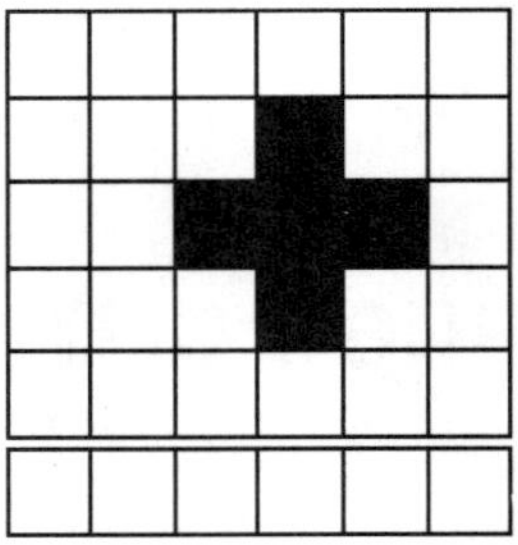

000000
000100
001110
000100
000000
000000

(Note: file will have all data sequentially as 000000000100001110000100000000000000

FIGURE 8.1 A BITMAP FORMAT IMAGE (LEFT) AND BITMAP FORMAT DATA.

This is ultimately how images are represented and viewed on monitors. A computer or television monitor has an addressable array of physical pixels, or dots. Each dot has position and color information.

Bitmap images are difficult to scale from their original resolution (without digital signal processing), and transporting them cumbersome because of their bulk. You can mitigate the size problem by using compression (covered soon) but only at the expense of increasing the time it takes to decode and render them.

Vector Formats

Now let's consider a way to represent images at a higher level of abstraction. Suppose you store endpoints of line segments that represent an image, perhaps to render a CAD drawing. You store the coordinates of the starting point, a direction, a length, and perhaps color information. The portion of the screen that does not have line segments will be rendered as a background color. The DiffCAD program uses a *vec* file format that stores line segments in an ASCII text file.

Vector formats typically store not only line primitives but also some 2-D shapes—such as circles, squares, and curved lines—that exist at higher levels of abstraction. Such shapes could be used, for example, to create jet planes or integrated circuit layouts. Figure 8.2 shows an example of a simple vector image and its associated data.

You could continue, using higher and higher levels of abstraction, including 3-D objects as primitives and avatars (3-D computer puppets) in virtual worlds.

One advantage of a vector format image is that you can scale the image easily without loss of detail. Many clipart collections are stored as vector format files for this reason. A disadvantage of vector formats is that it is hard to store highly detailed image information such as photographs, in which you may need to vary color information pixel by pixel.

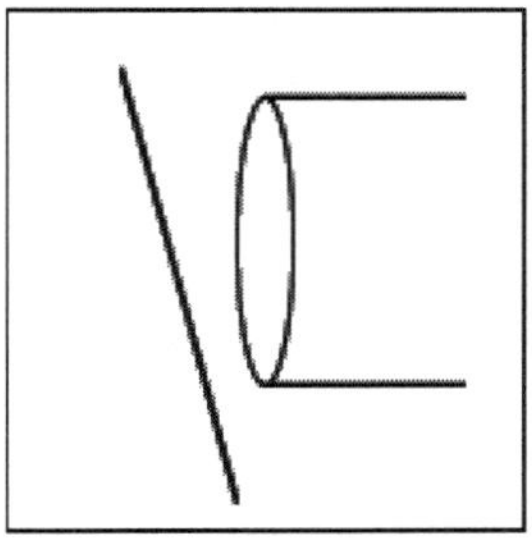

;ELLIPSE, 60, 50, 30, 45, BLACK; LINE, 40, 5, 30, 95, BLACK; LINE, 50, 80, 90, 80, BLACK; LINE, 50, 30, 90, 30, BLACK;

FIGURE 8.2 VECTOR FORMAT IMAGE (LEFT) AND VECTOR FORMAT DATA.

Conversion between Vector and Bitmap Formats

Converting an image from a vector format to a bitmap format is easy and straightforward; this conversion is common, because most display output devices are bitmapped. For a highly detailed vector format image, it is important to choose a sufficiently high resolution for the destination bitmap. Otherwise, artifacts—such as jagged lines (called jaggies) instead of straight lines—will appear in the image.

Converting an image from a bitmap format to a vector format is difficult. In Chapter 9, you will meet this formidable challenge with DSP routines for edge and outline detection. Another issue is the possible loss of color information when you convert an image from a rich bitmap representation to a (possibly) poor vector representation.

Other types of formats

[Murray et al.] describe several other types of formats for digital images.

- A *scene* format file has the condensed representation of an image. It is sometimes hard to tell the difference between this format and a vector format.
- A *metafile* can store both vector format elements and bitmap format elements. Examples of this type of format are PICT and CGM. Because of their versatility, these formats are often used to bridge the gap between hardware or software platforms.
- *Animation* formats come in many flavors. The simplest type stores adjacent frames of an animation sequence in one file for playing. Another type stores not only images but also color maps; changing the color map gives the illusion of motion. A more sophisticated animation

format stores frame difference information along with key frames. This technique, also used to store video, exploits the fact information typically changes little from frame to frame. Much of the background and features are static. If you store only the data that changes, you save a lot of space.

- *Multimedia* formats allow you to store various data types and formats together; video information, text information, and sound information can coexist peacefully.
- *Three-dimensional* formats support not only descriptions of lines, shapes, and 3-D geometries but also textures, reflections, and anything else a rendering program needs to reconstruct a 3-D image or world. Objects in a 3-D file are sometimes called *scene elements.* Many existing vector file formats have been extended to support 3-D. Such formats, such as Autodesk's DXF format, are referred to as *extended vector formats.* VRML is a little more than a 3-D format, because it includes support for HTML-style linking to other URLs on the World Wide Web.
- *Fonts* are special graphics files. They come in their own subsets of types (stroke, outline) based on bitmap or vector formats. The header or footer of a font file usually contains a database index associated with the font data.
- *Page Description Language* (PDL) formats are usually textual, programmatic descriptions of how to render graphics and text. An example is the ubiquitous PostScript format. This format is more akin to source code than to graphics data and requires a sophisticated program to create output.

Color Depth, Palettes and Transparency

Before an image is rendered on a monitor, regardless of the format it is stored in by an application, it is usually represented by a bitmap in a special memory known as a *frame buffer.* (The exception is images displayed on a random scan display monitor—a vector image display monitor.) The frame buffer stores the image as it is scanned by the graphics processor that feeds a CRT monitor for viewing. En route to the monitor, the graphics processor may be commanded by the host processor to manipulate the stream of pixels before they are displayed. One example of processing is to convert *color depth* information by using a *color lookup table* (CLUT), also known as a *color palette.* Color depth refers to the number of bits that are used to represent one pixel, or *bits per pixel* (bpp). For a color depth of 1, you represent a black and white image; a 1 turns on a pixel (white) and a 0 turns off a pixel (black). A color depth of 8 bpp means that you have 256 (2^8) possible values for a color.

A CLUT is a table of color values with an index, and the use of this table can allow for some image size reduction. You essentially form an indirect addressing scheme. Suppose you have 256 possible values (indexes) for a color. You can store 24-bpp color values in each entry of the CLUT, because your hardware supports it. You achieve some data compression, because the image data can reference 8-bit CLUT indexes instead of 24-bit color data. The image data must, however, also include the CLUT. The total data for a 640x480 image is (640x480x8 + 256x24)/8 = 308K. If the CLUT were not used, the total data would be (640x480x24)/8 = 922K. In this example, the use of a CLUT results in a savings of two-thirds in file size.

Using a CLUT doesn t always make sense. If an image contains a large number of colors, it may be more space-efficient to store the full pixel value directly. Generally, for images with more than 256 colors, it is better to use a *literal* or *absolute* format, because the overhead of a very large CLUT is not worth the space. In fact, the size of the CLUT may approach the size of the image.

Transparency

In the television world, live video is often overlaid onto a static image (such as a weather map). In the bitmap world, two bitmaps can be overlaid; portions of one bitmap can be specified as transparent to allow the background image to show through. Similarly, you could use transparency characteristics to create a fade from one video source to another. In this case, there would be degrees of transparency (not just on or off). Transparency is often described in bitmaps on a pixel-by-pixel basis. Here, transparency information is appended to pixel value information. The TGA format, for example, uses 5 bits each for R, G, B (red, green, blue) and 1 extra bit for transparency, for a total of 16 bits. When the transparency bit is on, the display hardware ignores that particular pixel so that any background image shows through. A 32-bit variant of the TGA format specifies 8 bits for transparency, called the *alpha channel.* Here, R, G, and B use 8 bits each, and alpha uses an additional 8 bits to specify the degree of transparency (0 = completely transparent; 255 = completely opaque).

Graphics Formats

This section gives you a directory of popular image formats. Table 8.1 (adapted from [Murray et al.]) shows a list of various formats along with their type. DiffCAD uses a subset of these formats: GIF, JPEG, VEC, PICT and

PPM. Later, we'll discuss general characteristics of graphics formats, such as file organization, compression, and progressive display.

TABLE 8.1 IMAGE FORMATS AND TYPE

Format	Type
AutoCAD DXF	Vector
Autodesk 3D Studio	Scene Description
BMP (Windows)	Bitmap
CGM	Metafile
FLI	Animation
GEM Raster	Bitmap
GEM VDI	Metafile
GIF	Bitmap
Harvard Graphics	Metafile
IFF	Bitmap
Intel DVI	Multimedia
JPEG File Interchange	Bitmap
Kodak PhotoCD	Bitmap
MPEG	Multimedia
PCX (Windows)	Bitmap
PICT (Mac)	Metafile
Pixar RIB	Scene Description
PNG	Bitmap
POV	Vector
PPM	Bitmap
QuickTime	Multimedia
Rayshade	Scene Description
SPIFF	Bitmap
Sun Raster	Bitmap
TIFF	Bitmap
TTDDD	Vector and Animation
Utah RLE	Bitmap
VEC	Vector
WMF (Windows)	Metafile

Most graphics formats support some form of data compression, discussed next.

Compression Methods: Making Bits of Bits

Some compression methods are used on digital information (and not just image data), and other methods are specifically suited for image data and other special classes of data. First, we'll look at four methods of general data compression: RLE, LZW, Huffman encoding, and arithmetic encoding. Later in this section we'll look at a compression method that is well suited for image data: Discrete Cosine Transform (DCT) or transform-based compression.

Compression can be *lossy* or can be perfect (*lossless*). Lossy compression methods discard information that is considered to be the least relevant in a particular application.

Compression can be *symmetric*—that is, the process of compression is similar in complexity, time, and methodology to decompression. On the other hand, in *asymmetric* compression a more complicated process is needed for one direction compared to the other. In the original Intel DVI video format, for example, a parallel supercomputer is used to compress a video sequence; a tiny amount of microcode in a video DSP chip is used to decompress the sequence. This method is highly asymmetric.

For bitmap format files, usually only the bitmap data is compressed. Any other information in the file (such as the header or footer) is left uncompressed for easy reading. For vector format files, there is usually no compression. Vector formats are inherently compact because they work at a higher level of abstraction than bitmaps. Also, rendering a vector format file takes a lot of time to begin with, and adding decompression would further slow applications that use vector format files.

NOTE

In the discussions that follow, the term encoding is used. Decoding is the inverse operation.

Run Length Encoding

Run length encoding (RLE) is a general compression method that encodes sequences of a character compactly as a number and the character.

For example:

```
AAAAAAAAAABC
```

could be coded as:

```
10A1B1C
```

The number 10 is the run count, and the letter A is the run value. If each ASCII character takes 1 byte of storage, the original uncompressed string takes 12 bytes, whereas the compressed string takes 7 bytes. Notice that even a run length of 1 requires a minimum of two characters.

For binary character encoding, there are several choices. You can encode on a bit basis (looking for runs of bits), on a byte basis, or on a pixel basis, where a pixel can take numerous bytes. The overhead of storing a run length for each run value can cause a file to be larger than the original, something that is termed *negative compression*. One method to minimize the effect of the run length code for small runs is to encode a bit at the beginning of each block that enables run length interpretation for that block. In other words, if the enable bit is set to 1, the block is interpreted as run length encoded. If the enable bit is set to 0, the ensuing data is interpreted as unencoded, or literal, data.

With 2-D bitmap data, you have the freedom to encode data along rows, which are referred to as *scanlines*, along columns, or along some other portion of the data

Lempel, Ziv, and Welch Compression

A widely used algorithm for data compression was invented by Lempel, Ziv, and Welch and is known as LZW. Actually, there are several algorithms. LZ77, LZ78, and LZW are all patented so their use may be subject to licensing fees and legal headaches. It is possible to adapt LZ77 so that you do not infringe on its patent (see the discussion of PKZIP later). Unfortunately, several widespread programs and formats, such as the Compuserve GIF file format, used these patented algorithms. In a backlash against having to pay for what used to be in the public domain, several alternatives to the patented LZW algorithms were developed and offered to the public. The popular archiving compressor PKZIP replaced the original LZW compressor with one based on an adapted, non-infringing variation of the LZ77 algorithm. GIF is encumbered by infringement problems, but the world is still full of GIF images. There are many freeware utilities to convert GIF files to other formats, such as PPM. The PNG graphics format was created as an alternative to GIF and is also based on a non-infringing variation of the LZ77 algorithm. DiffCAD avoids infringement of these patents by leaving the image processing features unimplemented. A programmer must alter the *lyon.ipl.license.java* file in order to implement the image processing feature.

How does LZW work? The LZ family of compressors are *dictionary-based* encoding algorithms. As data is read by a compressor, a table, or *data dictionary,* is built that has entries for patterns that occur in the input data stream. If new data is not in the dictionary, a new entry is made for it. When data is read that has a dictionary entry, the entry, which is smaller than the original data, is copied to the output (compressed) data stream. A key feature of LZW is that the dictionary need not be stored for the decoder; because of the way the data is organized, the decoder will be able to reconstruct the dictionary. This arrangement can save a lot of overhead and space. LZW is a lossless compression scheme.

Huffman Encoding

Like LZW, Huffman encoding is based on code words. Here, shorter codes are chosen to represent the most commonly occurring sequences in a data stream; longer codes are used for less frequent sequences. For example, the letter *A*, if used frequently in input text, may be coded with 2 bits instead of the usual ASCII 8 bits, whereas the letter *Q*, which occurs infrequently, may be coded with 12 bits. For the decoder to do its work, it must have the dictionary used for encoding. There is no on-the-fly construction of a dictionary as in the LZW compressor. The data stream that is produced by Huffman encoding has a subtle requirement: No code word can be the prefix of any other code word. In this way, a decoder can uniquely determine each entry of the table based on a sequential read of the data stream. An improvement on Huffman encoding is arithmetic encoding, which is discussed next. Both Huffman encoding and arithmetic encoding are lossless compression schemes.

Arithmetic Encoding

Arithmetic encoding, or *entropy coding,* improves on Huffman encoding in a couple of ways. First, you can have fractional codes—that is, you can have a 4.18-bit code (this is defined in a statistical way). Second, more-complex statistics are used. Context information is analyzed to derive a code for an input pattern. A *U* may be assigned a long code because it does not occur very often, whereas a *U* following a *Q* may be assigned a short code because a *U* is very likely to follow a *Q*. A brand of arithmetic encoding, called Q-coder is patented by IBM and AT&T and is subject to licensing considerations. An extension of the JPEG compression standard uses Q-coder.

DCT-Based Compression

The (DCT) method converts image data to the frequency domain. DCT is much like the DFT (Discrete Fourier Transform), which is discussed at length in Chapter 9. The DCT is a special case of the DFT [Netravali et al.]. The transform yields a set of values that correspond to magnitudes of frequency components. The human eye cannot distinguish very high frequency color changes, and this information can be discarded without much loss of detail in an image. Also, transform values that are zero or close to zero can be effectively compressed with a lossless compression scheme such as Huffman encoding, as can be done in JPEG. The overall JPEG compression method is lossy.

Progressive Display and the Internet

For users of the Internet and the World Wide Web, it is useful to enable progressive display of graphics images. When a user arrives at a new Web destination, any graphics data that is loaded for display is shown while it is loading. In this way, the user can immediately recognize the image instead of waiting for the entire image file to load before seeing anything. A few formats support progressive display: JPEG, GIF, and its patent-free successor, PNG. GIF has an option to store graphics data with every eighth line of data, then every fourth line, then every second line, and finally every line, for a total of four passes over the data. You can see a preview of an image with only one-eighth of the data file. This GIF storage option is called the *interlaced* option. The *non-interlaced* storage option stores rows sequentially and does not support progressive display (see Figure 8.3).

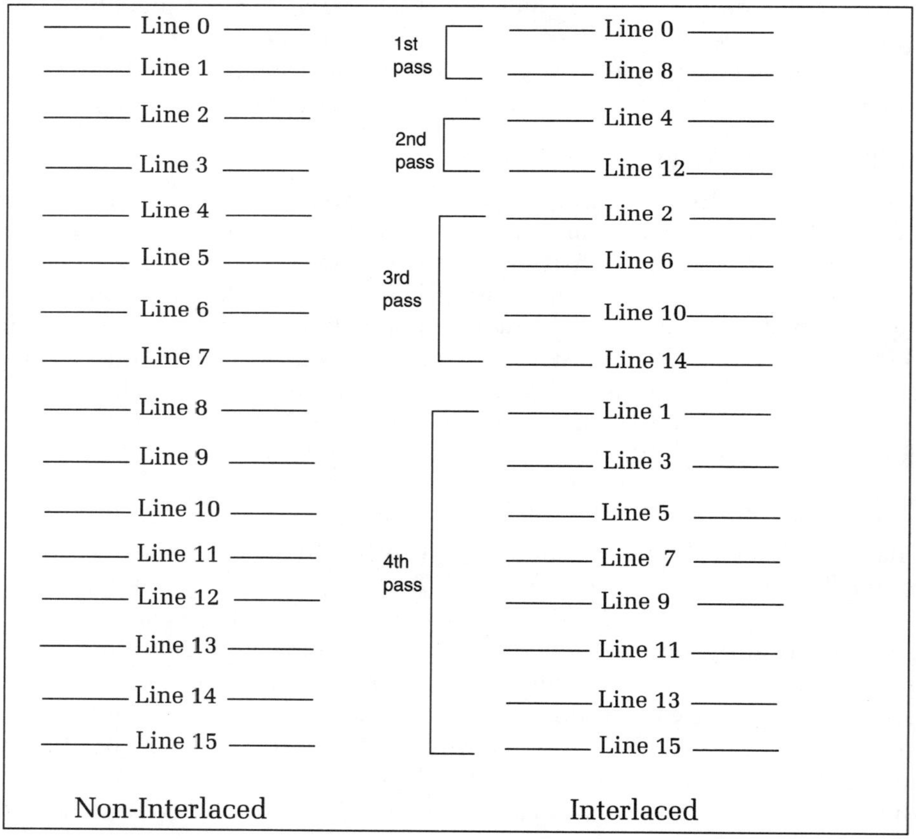

FIGURE 8.3 INTERLACED AND NON-INTERLACED STORAGE IN GIF.

The PNG format goes one up on GIF. PNG has an interlaced format, in which every eighth pixel of every eighth line is first transmitted. An image can be viewed with only 1/64th of the full image data.

JPEG data streams have an option for progressive display. Instead of scan-lines, the image is sent in progressively more-detailed layers. Approximations of the original image are sent in sequence so that the viewer sees the whole image right away; the quality of the image improves with time. Displaying each scan of progressive JPEG takes a full JPEG decompression cycle, and that can be CPU-intensive. Another extension of JPEG provides for hierarchical storage of the same image at multiple resolutions. A complete image is available at different resolutions to match the resolution of the display or print hardware.

Details of Several Formats

This section discusses in detail certain of the file formats, including those that are used by DiffCAD. Note that the Sun Java AWT class library provides built-in support for reading and writing JPEG and GIF image files. DiffCAD provides wrapper classes for some of this functionality. See the classes *WriteGIF*, *ReadGIF* and, *VSImage*.

GIF

GIF uses the LZW compressor. There are two versions: GIF87a (the original) and GIF89a. GIF89a may be incompatible with software that reads only GIF87a images, so most modern readers are expected to be able to read both formats. The formats are similar, but GIF89a has further extensions. Figure 8.4 shows the file layout for both formats, highlighting the differences.

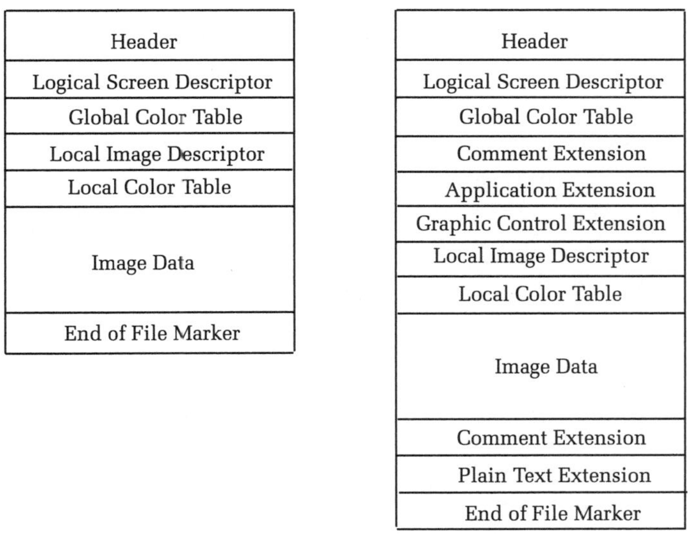

FIGURE 8.4 GIF87A (LEFT) AND GIF89A FILE LAYOUTS.

The file formats contain several pieces.

- Header. The header is six bytes in size. The first three bytes are "GIF" to identify the format; the next three bytes are the version ("87a" or "89a").
- Logical screen descriptor. This fixed-size group of bytes contains information about the minimum screen resolution (height and width) and color information to reproduce the image. If the screen is smaller than the screen parameters, the application will need to perform some scaling to display the image.
- Global color table. This optional section contains a CLUT of as many as 256 entries.
- Local image descriptor. This section has characteristics of the ensuing image data, including where on the display the image should start and the image resolution and color information.
- Local color table. GIF can be expanded to include more than one image, although this option is rarely used. A provision is therefore made to include a color table (termed "local") for each of the images. This is an optional table for specifying a CLUT for the image data that follows. This table, if present, supersedes the global color table.
- Image data. When compressed by LZW, image data usually comes out as a stream of data that must be read from beginning to end. GIF splits the data into a series of subblocks. Each subblock starts with a count byte, which can range from 1 to 255. The count byte value specifies the number of data bytes that will follow. A byte of value zero is used to terminate the subblock.

JPEG/JFIF

The JPEG standard contains ambiguities that make it an incomplete file format standard. C-Cube Microsystems created a file format called JPEG File Interchange Format (JFIF) that fills in the gaps. It is based on the baseline JPEG standard. JPEG is usually best applied to high-resolution, full-color (24-bpp) images, because the transform-based coding will have more latitude for compression with more color information. Keep in mind that sharp edges, such as those created by overlaid text, can become blurry. When compressing with JPEG, an application usually presents a quality setting that trades off compression for quality. To retain high-frequency detail in the source image, you can choose a high quality setting, although it will result in decreased compression. The trade-off between quality and compression is a thorny and persistent issue for JPEG.

A JPEG encoder uses the following steps:

1. Create header information.
2. Read the source image data in RGB.
3. Transform data to YUV color space. Y is black and white intensity information, and the U and V channels have color information. The transformation is linear (see Chapter 9 for color-space conversion).
4. Subsample the U and V channels. Discard any color information that is imperceptible to the viewer; use fewer samples of U and V for every sample of Y.
5. Perform the DCT on the Y, U, and V data.
6. Quantize the resulting coefficients into different bins. This process performs some compression by reducing the number of different possible values; more aggressive quantization is used for the color components.
7. Huffman-encode the quantized data and produce an output data stream.

Both a raw JPEG file and a JFIF file start with the bytes 255 and 232 to signify the start-of-image marker. For a JFIF file, you will see the bytes 255 and 240 followed by the characters "JFIF" and information about the image. Data that follows the first block is standard JPEG data as defined by the specification. For detailed information, obtain the specification from the American National Standards Institute [ANSI].

PPM

PPM is a bitmap format that is used as an intermediate format when you're converting from one system or file format to another. A set of portable freeware utilities written by Jeff Poskanzer converts to and from PPM to many other graphics file formats. For example, the ppmtogif utility converts from PPM to GIF.

The file organization for PPM is extremely simple. It begins with an ASCII header; the bitmap data follows as either ASCII or binary data. No compression is used. Data elements are separated by whitespace (a space, tab, carriage return, or linefeed).

The PPM header looks like the following:

```
MagicValue          P3= ASCII data, P6= binary data
ImageWidth          Width of image in pixels (ASCII decimal value)
ImageHeight         Height of image in pixels (ASCII decimal value)
MaxGrey             Maximum color value (ASCII decimal value)
```

The *MaxGrey* value specifies the maximum value for a color component. Each pixel is specified by three values for R, G, and B components.

Here is an example file:

```
# example of a 3 x 3 bitmap
P3
3 3
255
0 0 0            0 0 0            0 0 255
0 0 128  0 7 0            0 1  89
9 0  0           0 9  9           0  0  0
```

Comments can be included starting with the # character. The bitmap is 3 pixels high by 3 pixels wide. The third pixel of the second row has RGB values of

```
(0, 1, 89).
```

Because PPM is a simple format, here is the entire source code showing how DiffCAD implements a PPM reader.

```
/**
 * ReadPPM is a class that reads an image from
 * a PPM format file.
 *
 * Victor Silva (victor@cse.bridgeport.edu).
 *
 */


import java.io.*;
import java.awt.image.*;


public class ReadPPM
{
   public ReadPPM(InputStream in)
   {
   }


   private int type;
   private static final int PBM_ASCII = 1;
   private static final int PGM_ASCII = 2;
   private static final int PPM_ASCII = 3;
   private static final int PBM_RAW = 4;
   private static final int PGM_RAW = 5;
   private static final int PPM_RAW = 6;


   private int width = -1, height = -1;
   private int maxval;
```

```
   /// Subclasses implement this to read in enough of the image stream
   // to figure out the width and height.
   void readHeader(InputStream in) throws IOException
   {
      char c1, c2;

c1 = (char) readByte( in );
      c2 = (char) readByte( in );

if (c1 != 'P')
      {
         throw new IOException( "not a PBM/PGM/PPM file" );
      }
      switch(c2)
      {
      case '1':
         type = PBM_ASCII;
         break;

case '2':
         type = PGM_ASCII;
         break;

case '3':
         type = PPM_ASCII;
         break;

case '4':
         type = PBM_RAW;
         break;

case '5':
         type = PGM_RAW;
         break;

case '6':
         type = PPM_RAW;
         break;

default:
         throw new IOException( "not a standard PBM/PGM/PPM file" );
      }
      width = readInt( in );
      height = readInt( ir. );
      if ( type != PBM_ASCII && type != PBM_RAW )
```

```
      {
         maxval = readInt( in );
      }
   }

int getWidth()
   {
      return width;
   }

int getHeight()
   {
      return height;
   }

void readRow( InputStream in, int row, int[] rgbRow ) throws IOException
   {
      int col, r, g, b;
      int rgb = 0;
      char c;

for(col=0; col<width; col++)
      {
         switch(type)
         {
         case PBM_ASCII:
            c = readChar( in );
            if ( c == '1' )
            {
               rgb = 0xff000000;
            }
            else
            {
               if ( c == '0' )
               {
                  rgb = 0xffffffff;
               }
               else
               {
                  throw new IOException( "illegal PBM bit" );
               }
            }
            break;
```

```
            case PGM_ASCII:
               g = readInt(in);
               rgb = makeRgb(g, g, g);
               break;
            case PPM_ASCII:
               r = readInt( in );
               g = readInt( in );
               b = readInt( in );
               rgb = makeRgb( r, g, b );
               break;
            case PBM_RAW:
               if ( readBit( in ) )
               {
                  rgb = 0xff000000;
               }
               else
               {
                  rgb = 0xffffffff;
               }
               break;
            case PGM_RAW:
               g = readByte( in );
               if ( maxval != 255 )
               {
                  g = fixDepth( g );
               }
               rgb = makeRgb( g, g, g );
               break;
            case PPM_RAW:
               r = readByte( in );
               g = readByte( in );
               b = readByte( in );
               if ( maxval != 255 )
               {
                  r = fixDepth( r );
                  g = fixDepth( g );
                  b = fixDepth( b );
               }
               rgb = makeRgb( r, g, b );
               break;


default:
               break;
            }
            rgbRow[col] = rgb;
         }
      }
```

```
private static int readByte(InputStream in) throws IOException
   {
      int b = in.read();
```



```
// if end of file
      if (b == -1)
      {
         throw new EOFException();
      }
      return b;
   }


private int bitshift = -1;
   private int bits;


private boolean readBit( InputStream in ) throws IOException
   {
      if ( bitshift == -1 )
      {
         bits = readByte( in );
         bitshift = 7;
      }
      boolean bit = ( ( ( bits >> bitshift ) & 1 ) != 0 );
      —bitshift;
      return bit;
   }


/// Utility routine to read a character, ignoring comments.
   private static char readChar( InputStream in ) throws IOException
   {
      char c;


c = (char) readByte( in );
      if ( c == '#' )
      {
         do
         {
            c = (char) readByte( in );
         }
         while ( c != '\n' && c != '\r' );

      }
```

```
      return c;
   }

/// Utility routine to read the first non-whitespace character.
   private static char readNonwhiteChar( InputStream in ) throws IOException
   {
      char c;

      do
      {
         c = readChar( in );
      }
      while ( c == ' ' || c == '\t' || c == '\n' || c == '\r' );

      return c;
   }

/// Utility routine to read an ASCII integer, ignoring comments.
   private static int readInt( InputStream in ) throws IOException
   {
      char c;
      int i;

      c = readNonwhiteChar( in );
      if ( c < '0' || c > '9' )
      {
         throw new IOException( "junk in file where integer should be" );
      }

      i = 0;
      do
      {
         i = i * 10 + c - '0';
         c = readChar( in );
      }
      while ( c >= '0' && c <= '9' );

      return i;
   }
```

```
/// Utility routine to rescale a pixel value from a non-eight-bit maxval.
   private int fixDepth( int p )
   {
      return ( p * 255 + maxval / 2 ) / maxval;
   }


/// Utility routine make an RGBdefault pixel from three color values.
   private static int makeRgb( int r, int g, int b )
   {
      return 0xff000000 | ( r << 16 ) | ( g << 8 ) | b;
   }
}
```

VEC

The VEC format is a simple native vector file format used by DiffCAD. Data in the file is integer ASCII data. There are no other markers in the file. Two types are used in the format: POINT and VECTOR.

Points are stored as ASCII decimal numbers and are composed of two coordinates (x and y) that are separated by spaces or tabs. Points are separated by newlines (linefeeds). Here is a file describing a 3-pixel square.

```
0 0
3 0
3 3
0 3
```

Storing separate points can be useful for describe shapes and objects. Edge detection (see Chapter 9) of an image can create this sort of output. Also, this format can be used for computer vision and vector display applications.

The VECTOR type of data is specified in a similar manner except that two points are defined per line as follows:

```
x1 y1 x2 y2
x3 y3 x4 y4
...
```

Here, (x1, y1) defines the tail of a vector and (x2, y2) defines the head:

```
[(x1, y1), (2,y2)]
```

NOTE

DiffCAD has a routine to convert from xy points to vectors.

PICT

The PICT format can incorporate both bitmap and vector data. DiffCAD can write PICT vector data only. PICT is a fairly complex format, and few details will be shown here. PICT can use two different forms of compression: JPEG and PackBits. PackBits is an RLE-type encoding scheme.

Summary

Digital image file formats are defined by several characteristics: the type of file format (vector, bitmap, or other), the size efficiency based on the compression technology used, the number of colors handled and the supported resolution of images. Another factor useful for Internet-based graphics is progressive display, in which a partial image is shown as a graphics file is downloaded. Three formats have specific provisions for this option: GIF, PNG, and JPEG. Although the universe of graphics file formats is very large, there is a great deal of similarity between formats of the same type. One bitmap format is likely to be as capable as another. Metafile formats support vector and bitmap representations in a single format. Higher levels of abstraction are available in vector formats, which can represent 3-D objects, scenes, and worlds in 3-D formats. DiffCAD supports GIF, JPEG, PPM, VEC, and a subset of PICT.

Chapter 9

Image Processing in Java

The worth of a book is to be measured by what you can carry away from it.

—*James Bryce*

Save the mandrills, collect the whole set.

—*DL*

This chapter covers the computation of the histogram of an image. We follow this with a derivation of the basis for the 2-D Fast Fourier Transform (2-D FFT) and a summary of a class that implements the 2-D FFT. We show how to use the FFT to perform high- and low-pass filtering.

The high-pass filtering of the FFT is used to create edges, which are linked using a raster to vector converter. The raster to vector converter inputs an edge-detected image and outputs a series of line segments that can be drawn to the screen and saved as a PICT file.

We also cover color-space conversions and elementary 2-D rotation and scaling.

The Histogram

The *histogram* of an image is the probability mass function (PMF) of the pixel intensities. The PMF shows the statistical frequency of occurrence for an event. Thus, the PMF is computed by counting the number of times an event (a particular intensity) occurs in the data and then dividing by the total number of pixels in the image.

For example, suppose an image consists of a 1-D array given by the following:

$$\begin{bmatrix} 255 & 255 & 128 & 64 \end{bmatrix} \tag{9.1}$$

The PMF is computed, and the array is then expressed as an event with its associated PMF number:

$$\begin{bmatrix} event & PMF \\ 64 & 1/4 \\ 128 & 1/4 \\ 255 & 1/2 \end{bmatrix} \tag{9.2}$$

NOTE

The PMF is a discrete probability distribution function (PDF). Like the PDF, the PMF numbers will always sum to 1.

Naturally, the image arrays are much larger than that given in Equation 9.1. Furthermore, the array list shown in Equation 9.2 is typically shown as a bar chart. An example histogram is shown in Figure 9.1.

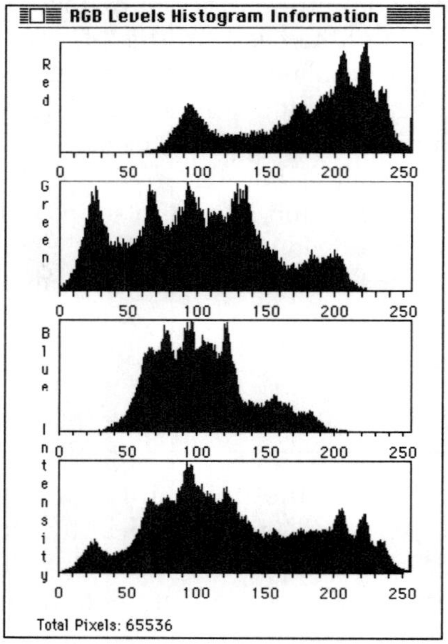

FIGURE 9.1 AN EXAMPLE HISTOGRAM.

The histogram of Figure 9.1 is shown for 255 intensities in red, green, blue (RGB), and intensity (I). The intensity is computed by averaging the RGB components and truncating to the nearest integer. The Histogram frame extends the *PictFrame* and, in doing so, saves the histogram as a PICT file (for editing). An example is shown in Figure 9.2.

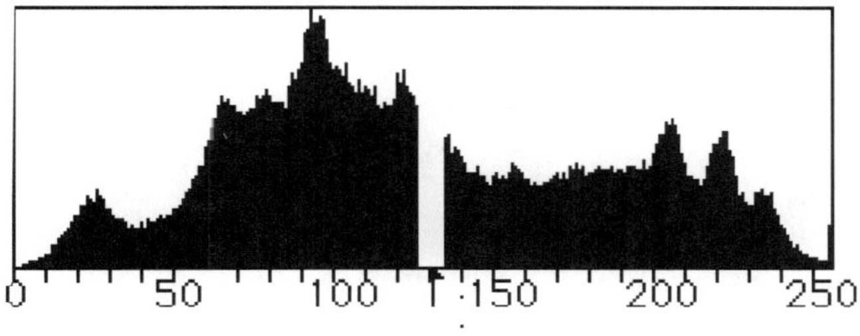

FIGURE 9.2 A SECTION OF THE HISTOGRAM FRAME'S PICT OUTPUT.

NOTE

Once the histogram is saved as PICT, the fonts can be changed so that they are no longer bitmapped. In this case, they are selected as Times Roman.

A code fragment from the *Histogram* class in the *lyon.ipl* package shows how to implement the display of Figure 9.1:

```
package lyon.ipl;
import java.awt.*;
import java.awt.image.*;
import gui.*;
import lyon.dclap.*;
class Histogram extends PicFrame {


    public int red[] = new int[256];
    public int green[] = new int[256];
    public int blue[] = new int[256];
    public int intensity[] = new int[256];
```

The constructor for the *Histogram* class takes a *PixelPlane* instance as an argument:

```
    public Histogram(PixelPlane p) {
...
```

Histogram uses the RGB values to act as indexes into three arrays of 256 integers.

```
for (int i=0; i<tp; i++) {
     red[p.getRed(i)]+=1;
     green[p.getGreen(i)]+=1;
     blue[p.getBlue(i)]+=1;
}
```

The *Histogram* constructor then computes the intensity by taking the truncated average of the three colors.

```
for (int i=0; i<intensity.length; i++) {
     intensity[i] = (red[i]+green[i]+blue[i])/3;
}
```

The rest of the code is devoted to normalization and display.

CD-ROM

The full source code for *Histogram* is available on the book's CD-ROM.

The 2-D DFT

Recall the 1-D DFT from Chapter 6:

$$V_k = \frac{1}{N}\sum_{j=0}^{N-1} e^{-2\pi ijk/N} v_j$$

The inverse DFT is given by the following:

$$v_j = \sum_{k=0}^{N-1} e^{2\pi ijk/N} V_k$$

Suppose that we have a 2-D array of uniformly sampled and quantized data:

$$f(x, y)$$

where

$$x \in [0..W-1], y \in [0..H-1]$$

The terms W and H are the width and height of the image in pixels.

The 2-D Discrete Fourier Transform (2-D DFT) is given by

$$F(u,v) = \frac{1}{WH}\sum_{x=0}^{W-1}\sum_{y=0}^{H-1} f(x,y)e^{-2\pi i(ux/W+vy/H)} \tag{9.3}$$

The inverse 2-D DFT (2-D IDFT) is given by

$$f(x,y) = \sum_{v=0}^{H-1}\sum_{u=0}^{W-1} F(u,v)e^{2\pi i(ux/W+vy/H)} \tag{9.4}$$

NOTE: The $(1/WH)$ term is not present in the IDFT, nor is the negative sign in the exponent.

We introduce a notation (following [Gonzalez et al.]) that uses a symbol, called the double arrow, to shorten the expressions in Equations 9.3 and 9.4 to the following:

$$f(x,y) \Leftrightarrow F(u,v) \tag{9.4a}$$

To turn the 2-D DFT into a 2-D FFT, we use the *separability* property to break the 2-D DFT into two fast 1-D FFTs. The exponential function's separability is due to the laws of exponents, namely, $e^a e^b = e^{(a+b)}$.

To employ the separability property, we factor Equation 9.3 into the following:

$$F(u,v) = \frac{1}{H}\sum_{y=0}^{H-1}\left[\frac{1}{W}\sum_{x=0}^{W-1} f(x,y)e^{-2\pi iux/W}\right]e^{-2\pi ivy/H} \tag{9.5}$$

Similarly, we factor the 2-D IDFT from Equation 9.4 into the following:

$$f(x,y) = \sum_{v=0}^{H-1}\left[\sum_{u=0}^{W-1} F(u,v)e^{2\pi iux/W}\right]e^{2\pi ivy/H} \tag{9.6}$$

The separability property means that the 2-D DFT can be computed by finding the 1-D DFT on the rows of the image and then finding the 1-D DFT on the columns. This means that we can use the 1-D DFT (and FFT) code from Chapter 6 to help perform the 2-D FFT.

Our implementation of Equation 9.5 transforms each row, placing the outcome in a complex array whose dimensions match those of the original image. Then we transform each column of the complex array. Thus, the 1-D DFT is performed on each row and again on each column.

Recall that the centering of Chapter 6 used

$$v_k = v_k(-1)^k$$

to cause a shift in the PSD. This comes about as a result of the time-shift theorem that states that in the frequency domain, spatial translation causes an added linear phase with slope proportional to the shift so that

$$f(x - x_0, y - y_0) \Leftrightarrow F(u, v)e^{-2\pi i(x_0 u/W + y_0 v/H)} \tag{9.7}$$

The dual of Equation 9.7 is

$$F(u - u_0, v - v_0) \Leftrightarrow f(x, y)e^{2\pi i(u_0 x/W + v_0 y/H)} \tag{9.8}$$

Where the left-hand side of Equation 9.7 is the Fourier transform of the right-hand side (using double arrow notation). Thus, a positional shift in the input plane causes a phase shift in the output plane. To put it another way, a positional shift in the time domain causes a phase shift in the frequency domain. This makes sense if we think of a sine wave being shifted in time. Relative to the unshifted sine wave, the shifted sine wave has a different phase. We represent a phase shift by multiplying by the complex exponential. To center the frequency on the 2-D DFT, we shift the frequency bins by $W/2$ and $H/2$. Thus, Equation 9.8 becomes

$$F(u - \frac{W}{2}, v - \frac{H}{2}) \Leftrightarrow f(x, y)e^{2\pi i(Wx/(2W) + Hy/(2H))} \tag{9.9}$$

This simplifies to

$$F(u - \frac{W}{2}, v - \frac{H}{2}) \Leftrightarrow f(x, y)e^{\pi i(x+y)} \tag{9.10}$$

By Euler's relation,

$$e^{i\theta} = \cos\theta + i\sin\theta$$

and because *x,y* are integers, we get the following:

$$e^{\pi i(x+y)} = \cos(\pi(x+y)) + i\sin(\pi(x+y)) = (-1)^{x+y} \tag{9.11}$$

Substituting Equation 9.11 into Equation 9.10 results in the following:

$$F(u-\frac{W}{2}, v-\frac{H}{2}) \Leftrightarrow f(x,y)(-1)^{x+y} \tag{9.12}$$

All 2-D FFTs (described in this book) are centered using Equation 9.12. A proof of the time-shift theorem for the 1D continuous case follows.

$$F[v(t-t_d)] = \int_{-\infty}^{\infty} v(t-t_d)e^{-i2\pi t}dt$$

let $\gamma = t - t_d$ and $dt = \gamma$ so that

$$F[v(t-t_d)] = \int_{-\infty}^{\infty} v(\gamma)e^{-i2\pi(\gamma+t_d)}d\gamma$$

so that

$$F[v(t-t_d)] = e^{-i2\pi t_d}\int_{-\infty}^{\infty} v(\gamma)e^{-i2\pi\gamma}d\gamma$$

Q.E.D.
We can take a similar approach in the 2-D discrete time domain to prove Equation 9.7:

$$f(x-x_0, y-y_0) \Leftrightarrow F(u,v)e^{-2\pi i(x_0 u/W + y_0 v/H)}$$

Invoking the definition of the continuous Fourier transform Equation 9.7 yields the following:

$$F(f(x-x_0, y-y_0)) = \iint f(x-x_0, y-y_0)e^{-2\pi i(ux/W+vy/H)}dxdy \tag{9.13}$$

Let

$$X = x - x_0 \text{ and } Y = y - y_0 \tag{9.14}$$

so that

$$dX = dx \text{ and } dY = dy$$

By substitution we obtain

$$F(f(x - x_0, y - y_0)) = \iint f(X,Y)e^{-2\pi i(u(x+x_0)/W+v(y+y_0)/H)}dXdY$$
$$F(f(x - x_0, y - y_0)) = e^{-2\pi i(ux_0/W+vy_0/H)}\iint f(X,Y)e^{-2\pi i(ux/W+vy/H)}dXdY$$

from which

$$f(x - x_0, y - y_0) \Leftrightarrow F(u,v)e^{-2\pi i(x_0u/W+y_0v/H)}$$

follows.
Q.E.D.
Further usage and optimization details are discussed in the following section.

The FFTPlane Class

The *FFTPlane* class resides in the *lyon.ipl* package. It provides an object-oriented 2-D color FFT and IFFT service. An instance of the *FFTPlane* class can be created from an instance of the *PixelPlane* class. *FFTPlane* treats all images as color and is not smart about conserving memory when working with achromatic images. *FFTPlane* makes a copy of the pixels in the *PixelPlane* instance. The copy is stored internally using complex red, green, and blue arrays of float type.

NOTE

It is important to set an *FFTPlane* instance to null to reclaim the memory used when it is finished.

FFTPlane has the ability to multiply each of its internally complex numbers in the frequency domain by real numbers stored in a *PixelPlane* instance. This arrangement enables the implementation of 2-D filters.

Class Summary

```
package lyon.ipl;
import java.awt.*;
import java.awt.image.*;
import java.io.*;
import gui.*;
import VS.*;
public class FFTPlane {
public FFTPlane(ProcessPlane ppIn)
public void mult( ProcessPlane ppIn)
public void fft()
public void ifft()
}
```

Class Usage

Suppose the following variables are predefined:

```
FFTPlane fftp;
ProcessPlane pp; // pp is a 2**n by 2**n image
ProcessPlane filter;
```

To make an instance of the *FFTPlane*, use the following:

```
fftp = new FFTPlane(pp);
```

NOTE

The *ProcessPlane* instance is altered by the methods in the *FFTPlane* instances.

To perform an in-place FFT on the instance of the *fftp*, use this scale:

```
fftp.fft();
```

The results of the *fft* method are left in internal data structures. The *ProcessPlane* instance is altered to show the log of the PSD.

To perform a multiplication, pixel by pixel, from the real pixels in *filter*, replacing the complex pixels in the *fftp* instance:

```
fftp.mult(filter);
```

NOTE

The *filter* instance should have the same dimensions as the *fftp*. Also, the pixels should vary from 0 to 255. They will be divided by 255 (so as not to increase the magnitude of the *fftp* result) before they are multiplied. The multiplication works with color filters (any *ProcessPlane* that contains colors).

To invoke an IFFT on an instance of the *FFTPlane* class and place the result in the original *ProcessPlane* instance, *pp* use:

```
fftp.ifft();
```

WARNING

This destroys the original *ProcessPlane* instance, *pp*.

In the following section, we show how to retrofit the *ProcessPlane* class so that any instance of *ProcessPlane* can support the *fft()*, *mult(pp)*, and *ifft()* invocations.

The ProcessPlane Implementation

We have retrofitted the *ProcessPlane* class with an *fft* method. We hold that this is the preferred way to perform a 2-D FFT, leaving the *FFTPlane* defined for those who would like the ability to extend its abilities. The *ProcessPlane* is retrofitted as follows:

```
     private FFTPlane fftp;
public void fft() {
     System.out.println("Running the FFT...");
     fftp = new FFTPlane(this);
     fftp.fft();
}
public void multFFT(ProcessPlane pp_) {
     if (fftp == null) {
          fft();
     }
     else {
          fftp.mult(pp_);
     }
}
public void ifft() {
     System.out.println("Running the iFFT...");
     if (fftp == null) {
          System.out.println("You must take the FFT first!");
     }
```

```
        else
            fftp.ifft();
}
```

The *ImageFrame* class has been updated to permit multiplication by images that are stored in the *childFrame*. The update for the *ImageFrame* class follows:

```
public void fft() {
        pp.fft();
        updateDisplay(pp);
}
public void multFFT() {
        pp.multFFT(childFrame.pp);
        updateDisplay(pp);
}

public void ifft() {
        pp.ifft();
        updateDisplay(pp);
}
```

DiffCAD and the Example 2-D FFTs

Some examples of Fourier transform pairs are shown in the following figures. On the left is *f(x,y)*. On the right is *F(u,v)*. The results are obtained using the DiffCAD program.

To improve the visibility of the PSD, we auto-scale and invert the PSD. Figure 9.3 shows one square painted with a bitmapped paint program called Debabelizer [Debabelizer].

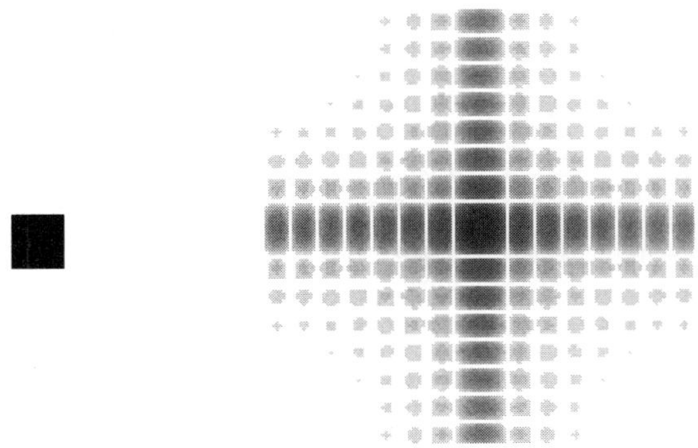

FIGURE 9.3 A SQUARE AND ITS PSD.

Figure 9.4 shows a black ring (which acts as a notch filter) with its PSD.

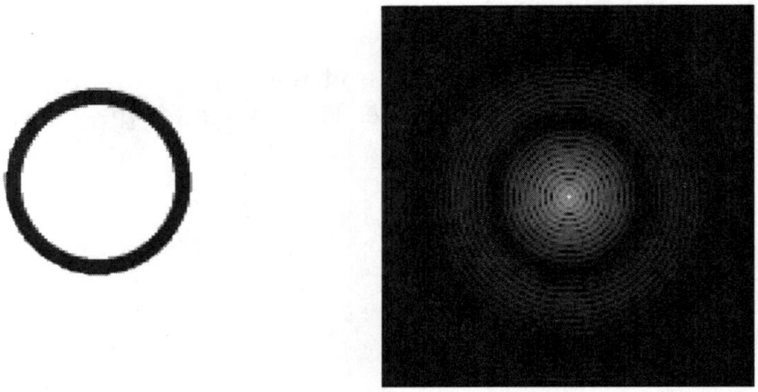

FIGURE 9.4 A BLACK RING AND ITS PSD.

These images are used as input. To gain a little more control than exists in most paint programs, we have retrofitted the *PixelPlane* class with a *dot* method. It permits the generation of a filled circle (like that shown in Figure 9.5) with a specified radius.

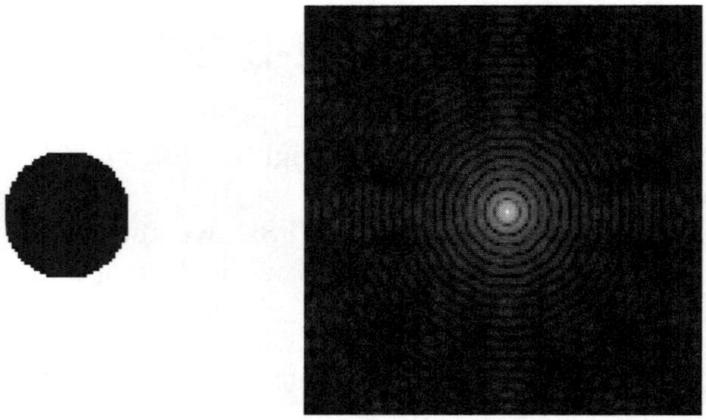

FIGURE 9.5 A BLACK DOT WITH ITS PSD.

The code for the *dot* method follows:

```
public void dot(int xc, int yc, int r) {


     int x1 = 0;
     int y1 = 0;
     int x2 = getWidth();
     int y2 = getHeight();
     float r2 = r * r;
     for (int x = x1; x < x2; x++)
          for (int y = y1; y < y2; y++) {
```

```
            if (((x-xc)*(x-xc) + (y-yc)*(y-yc)) < r2)
                setPixel(x,y,0,0,0,255);
            else
                setPixel(x,y,255,255,255,255);
        }
}
```

To synthesize diffraction gratings (a research goal of the DiffCAD program), we add a *lines* method to the *ProcessPlane* whose prototype is given by the following:

```
public void lines(Rectangle r, int increment)
```

Figure 9.6 shows an odd-shaped circle with its PSD.

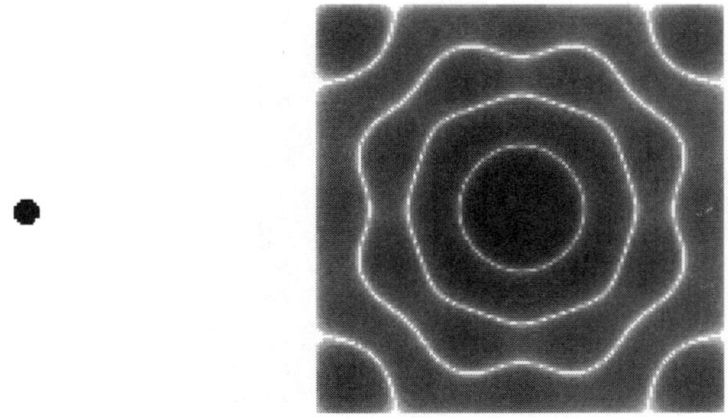

FIGURE 9.6 AN ODD-SHAPED CIRCLE WITH ITS PSD.

Figures 9.7–9.10 show PSDs multiplied by bilevel images. After the multiplication is performed, an IFFT is taken. This process enables the construction of images that can be used as filters in the frequency domain.

FIGURE 9.7 A TRIANGLE WITH ITS PSD.

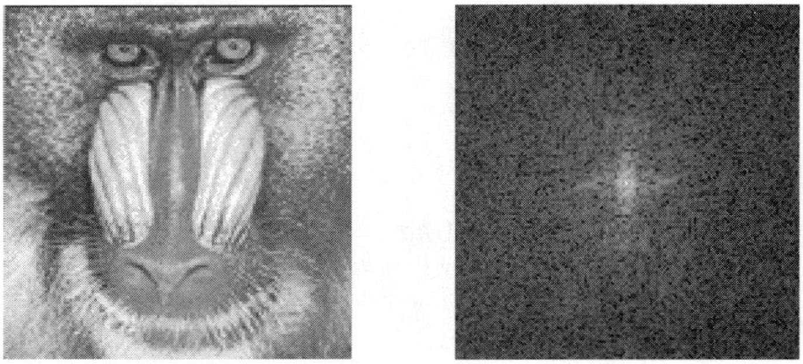

FIGURE 9.8 MANDRILL WITH ITS PSD.

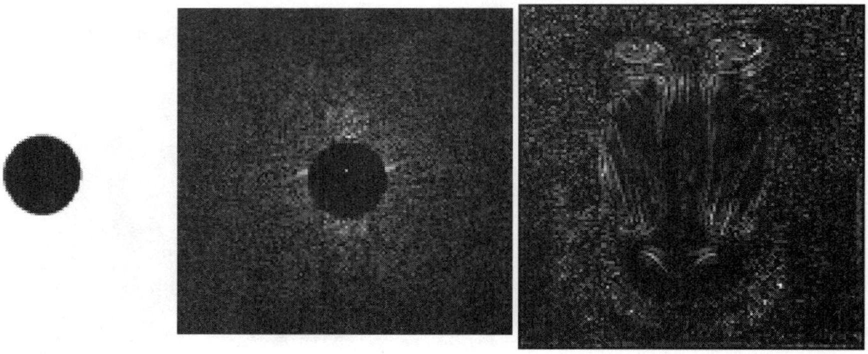

FIGURE 9.9 A FILTER THAT REMOVES SOME OF THE LOW FREQUENCIES FROM THE MANDRILL AND THE IFFT.

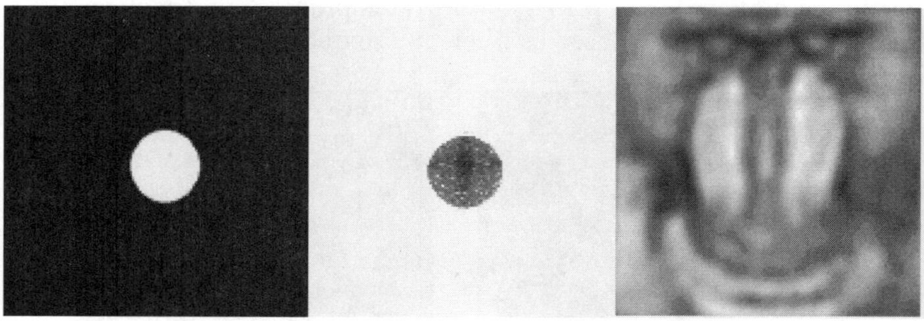

FIGURE 9.10 A FILTER THAT REMOVES SOME OF THE HIGH FREQUENCIES FROM THE MANDRILL AND THE IFFT.

Another neat trick we can perform with the FFT is Fraunhofer diffraction, a special case of Fresnel diffraction that occurs either when the distance from the grating (the aperture) is large or when the rules in the grating are close together. Fraunhofer diffraction is used when you're performing X-ray diffraction for studying underlying crystal-like structures (such as DNA). DiffCAD uses Fraunhofer diffraction to perform diffraction-based range finding [DeWitt and Lyon]. The following formula (whose derivation appears in [Walker]) shows the relationship between the DFT and Fraunhofer diffraction:

$$I(u,v,D) \approx \left| \frac{1}{\lambda D} \hat{A}\left(\frac{u}{\lambda D}, \frac{v}{\lambda D} \right) \right|^2 \tag{9.15}$$

Where (is the wavelength of light, $I(u,v,D)$is the intensity from Fraunhofer diffraction, and D is the distance between the target and the grating. (is the 2-D Fourier transform of the aperture (grating):

$$\hat{A}(u,v) = \iint_{\infty} A(x,y) e^{-2\pi i (ux+vy)} dydx \tag{9.16}$$

Figure 9.11, with its associated FFT, is an example of a far-field diffraction image computed by taking an FFT.

FIGURE 9.11 A GRATING WITH ASSOCIATED PSD.

An increment parameter is input using the scale dialog box and indicates the number of pixels to be skipped between lines. An example of the use of the *ProcessPlane* is given in the *ImageFrame* class:

```
public void lines() {
     pp.lines(bounds(),scaleKonst.getValue());
     updateDisplay(pp);
}
```

A variable-interrule-width grating (called a *chirp* grating) can give a radically different diffraction pattern from that of a fixed-width grating. An example is shown in Figure 9.12.

FIGURE 9.12 A CHIRP GRATING WITHPSD.

NOTE

For the diffraction images shown, $\lambda D = 1$. If $\lambda = 638$ nm (about the wavelength of a He-Ne laser) then $D = 1/\lambda$ =1.5 million meters.

Based on the assumption that the light is far, the FFT can assist us directly in the computation of Equation 9.16. At modest distance and with a small aperture relative to *D*, we can use the Fresnel diffraction formula:

$$I(u,v,D) \approx \left| \frac{1}{\lambda D} \iint_{\infty} A(x,y) e^{\frac{\pi i}{\lambda D}\left[(u-x)^2+(v-y)^2\right]} dydx \right|^2 \tag{9.17}$$

Developing code to solve Equation 9.17 is beyond the scope of this book. A numeric solution of Equation 9.17 can be found in [Baker].

Raster to Vector Conversion

Now let's see how to turn an edge-detected image into a list of line segments. We take an ad hoc approach to the conversion that runs in $O(N^2)$ time. For small numbers of *N* the quadratic growth does not appear to be a problem. We do not address refinements to the algorithm.

Converting a list of point coordinates into vectors (line segments) has applications in the graphic arts, CAD, and computer vision. In the graphic arts we use the vectors to form an editable outline of an image. The outline can be reproduced without the jaggies inherent in bitmapped images. In CAD we use the vectors to help construct a geometric model. In computer vision, we use the vectors to obtain geometric features that can be used in recognition. Another application for raster to vector conversion is in the output of line data using a vector-based device. For example, a pen plotter that uses a mechanical arm to deflect a pen must have the raster data converted to vector for efficient operation. Vector display systems (such as laser-based mirror deflection systems) form another class of devices that use the raster to vector conversion. In addition to converting from raster to vector, we must also order the vectors end-to-end to speed plotting on a mechanical pen plotter by minimizing the pen-up time. A pen plotter typically lifts a pen from the drawing surface before attempting to draw. A vector display device (such as an oscilloscope or laser) typically uses a blanking signal to turn off the beam. If the vector list is not retraced often enough, a human viewing the image will experience the sensation of flicker. Ordering the vectors maximizes the number of vectors to be displayed without flicker.

Traversing the vector list in the minimum time is a similar problem to that faced by a mail carrier starting from a post office and delivering letters to each block with minimum walking. This is called the *Chinese postman problem* [Roberts].

Another application of critical need in network communications is in data compression. If we could transmit a geometry faster than a bitmapped rendering of the geometry, we could render the geometry at any resolution using a Java program running on the client. For example, suppose that we wanted to distribute a popular test pattern (such as color bars). We could render the test pattern at various resolutions and then download them over the net, but this approach would waste of bandwidth and local data storage. Color bars consist of 11 rectangles of various sizes and colors. A 640 × 480 image of color bars could take more than 300K of memory. The program needed to store and display 11 rectangles takes less than 100 bytes of memory (and is much more flexible!).

Before a raster to vector conversion can start, we assume that we have good edges. Edge detection is a thorny problem and can be accomplished in one of several ways. We have already seen how to perform edge detection using a high-pass filter created with an FFT. In Chapter 7 we saw how a domain-specific edge detector can find the centroid of a thick edge. Sometimes we need to combine methods, such as the auto-scale and negate methods, with a threshold. In this section, we assume that a good edge-detected image is used as input. The goal is to take the edge-detected image and create a vector description of it. Figure 9.13 shows the output of the edge detector and the result of a raster to vector conversion. The file was saved as line segments to a PICT file.

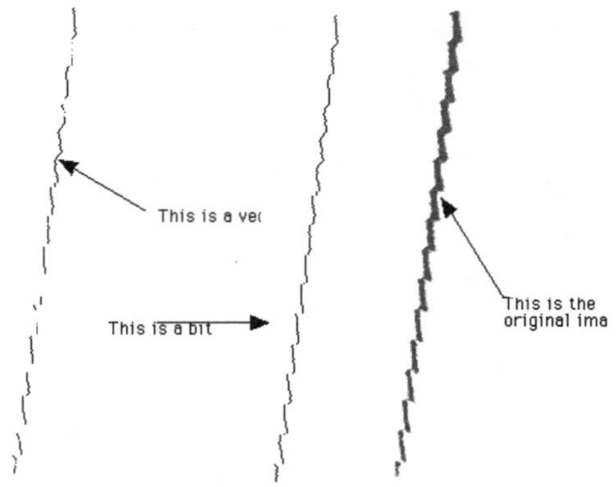

FIGURE 9.13 RASTER TO VECTOR CONVERTER.

The basic idea in raster to vector conversion is that a list of integer coordinates of the following form should be used as input:

```
x1 y1
x2 y2
...
```

The output consists of a series of vectors of the following form:

```
x1 y1 x2 y2
x3 y3 x4 y4
...
```

The idea is that the points used as input will exhibit intraframe coherence. We exploit this coherence by connecting the dots (the white pixels). The success of the algorithm depends on many points satisfying a *criterion of adjacency*, which is used to determine when two points are next to each other.

Typically, we say that two points are adjacent if they lie within a circle whose radius is one pixel. If two points are adjacent they can be used to form a line segment that is exactly two pixels long. The line segment consists of a head, a tail, and a slope. The compression ratio depends on the quality of the image edge detection, the amount of intraframe coherence, and the criterion of adjacency. Using a randomly selected GIF image found on the Internet, we were able to take 1,314 points (edge-detected pixels) and create 51 editable lines (a 25:1 compression ratio). When converting an image that contains lines (such as diffraction grating), the compression ratio increases to 132:1. Furthermore, our slope tolerance was very tight, so there was no introduced distortion. Our experience with various images indicates that it is difficult to generalize about the expected compression ratios.

A Raster to Vector Algorithm

In this section we present an algorithm that runs in $O(N^2)$, where N is the number of input points. We start with a set of points in a point list, *pl*. Our objective is to create a set of vectors, *v*. The following code is excerpted from the *Xy2vec* class in the *lyon.ipl* package:

```
public static void main(String args[]) {
Vector v = new Vector();
PointList pl = new PointList();
Xy2vec x = new Xy2vec();
```

The following line reads the points from an input file and places the data into the point list *pl*:

```
x.readPoints(pl);
```

The point list is treated as a stack, where the *popPoint* method returns a point from the top of the stack:

```
Points p = pl.popPoint();
```

Points is a point list data type, a class in the *lyon.ipl* package that we cover in a later section. In the following *while* loop, we check to make sure that the point list is not empty. Our objective is to place the point in one of a list of vectors. If we cannot, we make a new vector:

```
while (p != null) {
boolean point_not_stashed = true;
// Empty vector ?
if (v.size() == 0) {
```

The *Vec* class is another class in the *lyon.ipl* package. An instance of a *Vec* consists of a head, a tail, and a slope. When a *Vec* instance is first created, the tail is null. To keep track of this, we provide a flag called *tail_is_empty*. We set this flag to true whenever we make a new *Vec* instance. To complicate matters somewhat, the list of *Vec* instances is stored in an instance of the *Vector* class, *v*.

```
    Vec W = new Vec(p);
    W.tail_is_empty = true;
    v.addElement(W);
    point_not_stashed = false;
} else {
```

For each *Vec* instance in the list of vectors, *v*, we test to see whether the point is adjacent to the *Vec* instance. If it is, we test to see whether the slope is within a tolerance. If the point is adjacent to the vector's endpoints and the slope is within tolerance, we grow the vector by one point. We then declare that the point is placed in the list of vectors and move to the next point. If we cannot place the point, we create a new vector with the orphaned point at the head.

```
    for (int i=0; (i<(int)v.size())&&point_not_stashed;i++) {
        Vec newV = (Vec)v.elementAt(i);
        if (newV.head.isAdjacent(p)&&
        newV.isSlopeAcceptable(p)) {
        newV.addHead(p);
    point_not_stashed = false;
}
// Put new point to tail
if (point_not_stashed) {
    if ((newV.tail_is_empty &&
        newV.head.isAdjacent(p))||
        (!newV.tail_is_empty &&
        newV.tail.isAdjacent(p)&&
        newV.isSlopeAcceptable(p))) {
    newV.addTail(p);
    if (newV.tail_is_empty) {
        newV.tail_is_empty = false;
        newV.getSlope();}
    point_not_stashed = false;
    }


}
}// for
// Put new point to head
if (point_not_stashed) {
    Vec nv = new Vec(p);
nv.tail_is_empty = true;
v.addElement(nv);
point_not_stashed = false;
            }
    }
```

After we are finished with the point, we proceed to the next point:

```
    p = pl.popPoint(); // Get new point
}// while
x.printOutData( v);
```

We say that the algorithm is $O(N^2)$, because, in the worst case, we will be able to create vectors that are one pixel long. This is a pathological case created by an advisory and occurs when there is no pixel adjacency. This seldom happens. In fact, it might be better to consider an average case of $O(V^2)$, where V would be the expected number of vectors. Computing the average case is probably practical only when the domain is known in advance (such as in a compression ratio for text images).

The Slope Class

Two parameters are central in controlling the behavior of the raster to vector algorithm. The first, the slope tolerance, is stored in the *Slope* class. The second, the radius squared of a circle about the vector endpoints, is described in the next section. The following code resides in the *lyon.ipl* package and is contained in the *Slope.java* file:

```
package lyon.ipl;
class Slope {
public double dx, dy;
```

The slope of a vector is represented by the change in *y* divided by the change in *x*. As we attempt to extend an existing vector, we will add to a vector's endpoint only if the slope changes by less than some ε amount. The value of ε will depend on the desired compression ratio and the tolerance for loss in the compression. Lower values of ε will result in a higher compression ratio with greater loss.

```
public static double eps = 0.001;


Slope() {};


Slope(double dx_, double dy_) {
          dx = dx_;
          dy = dy_;
}
```

The *isEqual* method checks for ε difference in the slopes between the existing slope and a proposed slope.

```
public boolean isEqual(Slope s) {
     if (s.dx == 0 && dx == 0)
         return true;
    if ((s.dx != 0) &&( dx != 0))
        return (Math.abs((s.dy/s.dx) - (dy/dx)) < eps);
    else
        return false;
    }
} // class Slope
```

The Points Class

The *Points* class resides in the *lyon.ipl* package and is used to store a list of points that are used as input to the *Xy2vec* algorithm. A central parameter that controls the loss in the output as well as the compression ratio is the radius of a circle about a vector endpoint that determines adjacency.

To determine whether two points are adjacent, we use a Boolean method called *isAdjacent*, which we embed in the *Points* class:

```
package lyon.ipl;
class Points {
double x,y;
```

The square of the radius of the distance between two points that are judged adjacent is called *dr*.

```
double dr = 1.0;
Points(double px, double py) {
          x = px;
          y = py;
}
```

To decide whether two points are adjacent, we check the distance between them in pixels. This distance is compared with *dr*. We can adjust *dr* to be a number greater than 1, but it will introduce some geometric distortion:

```
 public boolean isAdjacent(Points p) {
    double dx = x-p.x;
    double dy = y - p.y;
    double r = dx * dx + dy * dy;

return r <= dr;
} // isPointsAdjacent


} // class Points
```

Color Models

Light energy is a form of electromagnetic radiation. As Maxwell said, it consists of waves propagated through an electromagnetic field according to electromagnetic laws [Banerjee]. A light source may radiate energy with several spectral components. To determine the spectra of a light source, an instrument called a spectrometer is used. A typical light source (such as a tungsten filament in an incandescent lamp) is heated to a temperature that causes a phenomenon known as *radiancy*. Radiancy is defined as the rate per unit surface area at which energy is radiated into the forward hemisphere [Resnick]. Radiancy is typically given in watts per centimeter squared and is found by integrating the *spectral radiancy*. The spectral radiancy is found by using a spectrometer to measure the amount of energy that falls over an area at a particular wavelength.

A *cavity radiator* is an idealized heated solid that emits a radiancy that is given by the following:

$$R_c = \sigma T^4 \tag{9.18}$$

Where T is in degrees Kelvin and maoer is the Stefan-Boltzmann constant. A typical cavity radiator is made of a block of metal (such as tantalum, tungsten, or molybdenum). The cavity is created by drilling a small hole. Radiancy is typically measured when temperatures are in the 1,500-7,000 K range. It is important to realize that the radiancy at the surface of various materials is a material property but that the radiancy of all cavity radiators is governed by Equation 9.18 and is independent of the material.

Another attribute of the cavity radiator (which is also known as a *blackbody radiator, full radiator,* or *Planckian radiator*) is the spectral radiancy. Spectral radiancy for the Planckian radiator is given by

$$M_e = \frac{c_1}{(e^{c_2/\lambda T} - 1)\lambda^5} \tag{9.19}$$

where

$$c_1 = 3.74183x10^{-16} Wm^2, \text{ and } c_2 = 1.4388x10^{-2} mK$$

[Hunt].

NOTE

All cavity radiators have the same spectral radiancy.

To summarize, light is energy that has a spectral radiancy. We can compute the PSD for light just as we do for sound. To reconstruct the light we must therefore duplicate the spectral radiancy. Such an effort may be impractical if the objective is to build a computer display for the purpose of human vision. For the rest of this section, we discuss how to reconstruct the spectral radiancy of light for the purpose of human perception. This assumption enables the construction of practical computer display systems.

The human perception of the spectral radiancy of light is called *color vision.* For the purpose of this discussion, we use the term *color* to mean the human perception of color. Our goal is to describe a color model that can assist in reconstructing colors for humans to see. In that case, color becomes a *psychophysical* quantity. The known relationships between the psychological color experience and physical light stimulation have given rise to several conflicting theories about how the human nervous system works [Teevan].

Color sensation in the human eye is produced when light of various wavelengths falls upon the eye. Most of the color models in common use are based on the *tristimulus* theory of color perception. The theory is based on a physiological approach that attempts to explain the eye's behavior in terms of its components. The eye is a photosensitive sensor that contains optical elements and photoreceptor cells. There are two types of photoreceptor cells: rods and cones. These cells enable a spectral response for human vision, which varies not only from person to person but also as an individual ages. Age alters the spectral response of the human eye, because age causes a progressive yellowing of the lens of the eye. Thus, there is no general agreement on the spectral sensitivities of the human visual system [Cowan et al.].

Rods are responsible for low-light imaging and do not play a part in normal color vision. The rods have varying amounts of a photosensitive pigment, called *rhodopsin,* that absorbs light most strongly in the blue-green part of the spectrum.

Cones are divided into three types: R, G, and B. It has not yet been possible to isolate the pigments in cones [Hunt]. The cones have been found to have a logarithmic response. Also, the spectral response for each of the cones differs in central frequency, peak amplitude, and bandwidth [Faugeras].

The tristimulus theory of color perception says that we need only three primary colors to create the gamut of visible color. Also, an examination of the spectral response curves should enable us to approximate the colors for which the cones have a peak spectral response. These curves happen at about 444 nm (B), 526 nm (G), and 645 nm (R) [Cohen et al.].

NOTE

The red and green cones have a similar spectral sensitivity.

The typical range in human vision is from 380 to 770 nm. So RGB monitors (and systems for the storage of image data) have their roots in the human visual system. Furthermore, the perceptual system is logarithmic in response (for both vision and hearing!). This accounts for the popularity of the logarithmic companding techniques discussed in Chapter 5.

The tristimulus theory leads to a color model that consists of a 3-D color space. A color model provides a framework for color specification. There are two kinds of color synthesis: additive synthesis and subtractive synthesis.

The *additive synthesis* color model adds light created via radiance. The RGB color model is a common example of an additive synthesis color model. In the RGB color model, red, green, and blue light are added in equal amounts to create various shades of gray. When red, green, and blue light are combined in equal amounts, at maximum intensity (on, for example, a computer monitor) the color is called white.

The *subtractive synthesis* color model removes light energy via absorption. The cyan, yellow, and magenta pigments are applied in equal amounts to make varying shades of black. Figure 9.14 shows the relationship between the RGB system and the CMY system. When colors are combined in equal amounts, the color-space points lie on the main diagonal between the white and black points on the color cube.

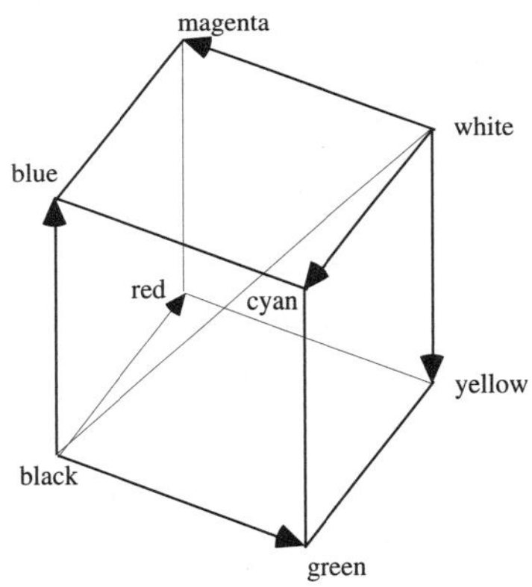

FIGURE 9.14 ADDITIVE VS. SUBTRACTIVE COLOR SYNTHESIS.

The HLS System

One of the color spaces held in common use is the hue, luminance, and saturation (HLS) model. This model uses a cylindrical coordinate system, in which the magnitude of a vector is specified as a function of the height and angle. The height is used to represent luminance, and the angle is used to represent hue. The magnitude of the vector represents the saturation. The basic idea of the conversion is that red is assigned a hue angle of 0 degrees, blue, 240 degrees, and green, 120 degrees. Luminance and saturation vary from zero to 1, and the hue varies from zero to 360 degrees.

The *ImageFrame* class has been modified to perform the RGB to HLS and HLS to RGB conversions. These facilities are built into the *ProcessPlane* (*pp*) instance. Their invocation follows:

```
public void rgb2hls() {
     pp.rgb2hls();
     updateDisplay(pp);
}
public void hls2rgb() {
     pp.hls2rgb();
     updateDisplay(pp);
}
```

These are methods in the *ImageFrame*. The *ProcessPlane* instance is stored in *pp*. Menu items have been added to the *ImageFrame* to automate the invocation of the conversions. Figure 9.15 shows the Color menu from the *ImageFrame*.

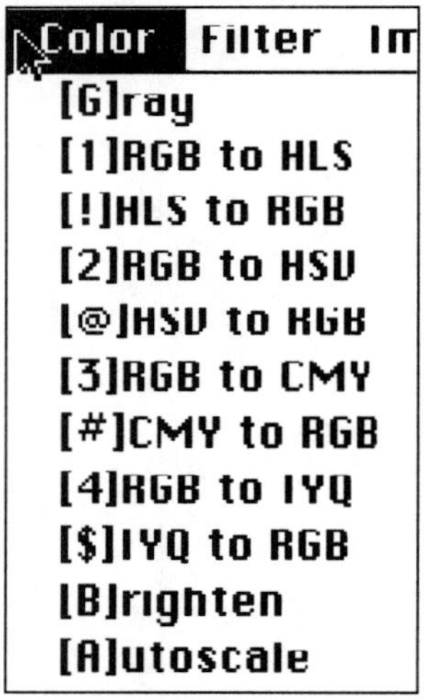

FIGURE 9.15 THE COLOR MENU IN THE IMAGEFRAME

One complicating aspect of color-space computation is the use of floating-point numbers in the transformation matrices. Recall that in the *ProcessPlane* class, we store an image as a packed array of 32-bit ints. Such an array is unsuitable for the storage of floating-point computations, because round-off error will build rapidly when only 8 bits are used to represent each color. To mitigate the effect of round-off error and to facilitate other floating-point computations, we have elected to store the images as three arrays of float. This approach presents the programmer with the burden of setting instances of the color conversion classes to null in order to reclaim memory. It also requires that the garbage collector be explicitly invoked if out-of-memory errors occur. We have found that this invocation is seldom needed, because we have associated these instances with windows that are under the users' control. When a user disposes of an *ImageFrame* instance, all image data is reclaimed. Thus, memory errors will occur if too many windows are open. As a result, the user must close some windows before proceeding.

The IYQ System

The IYQ system was invented for color television transmission. The symbol *I* stands for in-phase, *Y* for luminance, and *Q* for quadrature. The *IYQ* color space is used in several analog color television transmission systems, including PAL (Phase Alternating Lines), NTSC (National Television Systems Committee), and SECAM (Sequentiel Couleur avec Memoire). NTSC is used in the United States, Canada, Mexico, and Japan. SECAM is used in France, the former USSR, and eastern Europe. PAL is used in western Europe.

The IYQ system is based on the idea that the human visual system requires crisp outlines, but that it can tolerate lower color bandwidths. This system reduces color bandwidth by low-pass filtering the I and Q color components. Further, because the eye is less sensitive to magenta than to orange, the I color component (orange-cyan) is given more bandwidth than the Q color component (green-magenta). Thus there are compelling human visual reasons for using the IYQ system (even in non-analog video compression systems) [Inglis].

$$\begin{bmatrix} Y \\ I \\ Q \end{bmatrix} = \begin{bmatrix} 0.299 & 0.587 & 0.114 \\ 0.596 & -0.274 & -0.322 \\ 0.211 & -0.522 & 0.311 \end{bmatrix} \begin{bmatrix} R \\ G \\ B \end{bmatrix} \tag{9.20}$$

To convert from the IYQ space, we use the following:

$$\begin{bmatrix} R \\ G \\ B \end{bmatrix} = \begin{bmatrix} 0.299 & 0.587 & 0.114 \\ 0.596 & -0.274 & -0.322 \\ 0.211 & -0.522 & 0.311 \end{bmatrix}^{-1} \begin{bmatrix} Y \\ I \\ Q \end{bmatrix} \tag{9.21}$$

For reasons of computational precision, we do not evaluate the matrix inverse expressed in Equation 9.21. Some books give the matrix and its inverse with three significant figures of resolution [Rogers]. We can get a much more precise answer if we allow Java to perform the inverse using the precision of a float or a double. This approach is particularly important, because we will use a floating-point format to store the pixels.

Some images (such as X-rays, MRI, and Hubble telescope pictures) contain more than 8 bits of color information per color plane. A floating-point format is ideal for storing this type of image. Although display buffers show only 8 bits per color, we need not introduce that limitation into our computations or data files.

The FloatImage Class

The *FloatImage* class resides in the *lyon.ipl* package. Its purpose is to turn a *ProcessPlane* instance into three arrays of floats, which are needed to perform high-precision image processing. A reference to the *ProcessPlane* instance that is used to create the *FloatImage* is kept in internal private storage.

Class Summary

```
package lyon.ipl;
import java.net.URL;
import java.applet.Applet;
import java.awt.*;
import java.awt.image.*;
import futils.utils.*;


public class FloatImage {
  public float r[];
  public float g[];
  public float b[];
  public FloatImage(int l)
  public FloatImage(ProcessPlane pp_)
  public ProcessPlane makeProcessPlane()
  public int getLength()
  public int getHeight()
  public int getWidth()
  public float getRed(int i)
  public float getGreen(int i)
  public float getBlue(int i)
  public float getAlpha(int i)
  public float getRed(int x, int y)
  public float getGreen(int x, int y)
  public float getBlue(int x, int y)
  public float getAlpha(int x, int y)
```

```
  public void setPixel(int x, int y, float r_, float g_, float b_)
  public void setPixel(int i, float r_, float g_, float b_)
  public void printSize()
  public float max(int i)
  public float min(int i)
}
```

Class Usage

Suppose the following variables are predefined:

```
FloatImage fi;
ProcessPlane pp;
double red[];
double green[];
double blue[];
int l;
```

To get and set the color components from a *FloatImage* instance, we use this code:

```
red = fi.r;
green = fi.g;
blue = fi.b;
```

NOTE

After great deliberation, we have decided to make the RGB data structures public. The alternative is to fill code with *getArray* accessor methods. We are still not sure whether this was the right design choice. Typically, accessor methods are used to keep threads synchronized, but we have abandoned this approach. The resulting code is unsuitable for multithreading but is smaller, faster, and easier to understand.

To make an instance of a *FloatImage* with *l* pixels:

```
fi = new FloatImage(l);
```

To make an instance of a *FloatImage* by making a copy of a *ProcessPlane* instance:

```
fi = new FloatImage(pp);
```

NOTE

Operations performed on the *FloatImage* instance do not alter the data of the *ProcessPlane* instances. All elements in the *ProcessPlane* instance were copied and type-converted during the copy process.

To turn an instance of a *FloatImage* into an instance of a *ProcessPlane*:

```
pp = fi.makeProcessPlane();
```

To get the number of pixels in the *FloatImage* instance:

```
l = fi.getLength();
```

To get the height and width of the *FloatImage* instance:

```
int h = fi.getHeight();
int w = fi.getWidth();
```

To get the red, green, blue, and alpha components in the *FloatImage* instance, treating all the arrays as 1-D arrays:

```
int i;
float r = fi.getRed(i);
float g = fi.getGreen(i);
float b = fi.getBlue(i);
float a = fi.getAlpha(i);
```

NOTE

The alpha channel is never kept in the *FloatImage* instance. Instead, the alpha channel is converted from the *ProcessPlane* on demand. The reason is that no *FloatImage* methods alter the alpha channel, so no floating-point version of the alpha channel is needed.

To get the red, green, blue, and alpha components in the *FloatImage* instance, treating all the arrays as 2-D arrays:

```
int x, y;
float r = fi.getRed(x, y);
float g = fi.getGreen(x, y);
float b = fi.getBlue(x, y);
float a = fi.getAlpha(x, y);
```

To set a floating-point pixel using 2-D coordinates:

```
float r, g, b;
int x, y;
fi.setPixel(x, y, r, g, b);
```

To set a floating-point pixel using the 1-D internal order:

```
fi.setPixel(i, r, g, b);
```

To print the size of the array to *System.out*:

```
fi.printSize();
```

To find the minimum- and maximum-amplitude color components located at position *i* in the 1-D internal order:

```
float intensity = fi.max(i);
float intensity = fi.min(i);
```

The ColorConverter Class

The *ColorConverter* class is an abstract class that resides in the *VS* package. The *ColorConverter* class is extended with implementations required for methods that convert from and to the RGB color space. Internally, *ColorConverter* provides a storage area for the *FloatImage* and *ProcessPlane* instances.

Class Summary

```
package VS;
import java.awt.*;
import java.io.*;
import lyon.ipl.*;
public abstract class ColorConverter {
  public ProcessPlane pp;
  public FloatImage fi;
  public ColorConverter(ProcessPlane pp_)
  public abstract int[] fromRGB();
  public abstract int[] toRGB();
  public FloatImage getFloatImage()
}
```

Class Usage

A class that extends *ColorConverter* must implement the methods *fromRGB()* and *toRGB()*. A class that extends *ColorConverter* is used to convert a *ProcessPlane* instance (an RGB image) into an another color-space-encoded *FloatImage* instance. The *IYQ* class is an example of a class that extends *ColorConverter*:

```
package lyon.ipl;
import VS.*;
public class IYQ extends ColorConverter {
```

Recall that the RGB to IYQ conversion requires a matrix multiplication given by Equation 9.20:

$$\begin{bmatrix} Y \\ I \\ Q \end{bmatrix} = \begin{bmatrix} 0.299 & 0.587 & 0.114 \\ 0.596 & -0.274 & -0.322 \\ 0.211 & -0.522 & 0.311 \end{bmatrix} \begin{bmatrix} R \\ G \\ B \end{bmatrix}$$

An implementation of Equation 9.20 follows:

```
double A[][] = {
     { 0.299, 0.587,  0.114},
     { 0.596,-0.274, -0.322},
     { 0.211, 0.522,  0.311}
     };
```

To create instances of a matrix that knows how to invert itself and multiply itself by other matrices we use the *Mat3* class (discussed in the following section).

```
Mat3  rgb2iyqMat = new Mat3(A);
```

The matrix inversion in Java is performed at instantiation. The results are much more precise than those found in most publications:

```
Mat3  iyq2rgbMat = rgb2iyqMat.invert();
```

The constructor calls the superclass, *ColorConverter*, which copies the *ProcessPlane* instance into a new *FloatImage* instance. This constructor is required of any class that is to extend *ColorConverter*.

```
public IYQ ( ProcessPlane pp_) {
     super(pp_);
}
```

The following two methods are typical of any that implement a color-space conversion using a 3x3 matrix multiplication. This arrangement is common with color-space conversions, but several of them, such as polar coordinate color systems like HVS and HLS, do not perform the matrix multiplication of Equation 9.20.

```
public int[] fromRGB() {
double pel[];
float r, g, b;
      for (int i=0; i < fi.getLength(); i++) {
           pel = rgb2iyqMat.multiply(fi.r[i],fi.g[i],fi.b[i]);
        fi.r[i] = (float) pel[0];
        fi.g[i] = (float) pel[1];
        fi.b[i] = (float) pel[2];
        pp.setPixel(i, (int) pel[0], (int) pel[1], (int)pel[2], 255);
      }
      return(pp.pels);
}
public int[] toRGB() {
double pel[];
      for (int i=0; i < fi.r.length; i++) {
           pel = iyq2rgbMat.multiply(fi.r[i],fi.g[i], fi.b[i]);
        fi.r[i] = (float) pel[0];
        fi.g[i] = (float) pel[1];
        fi.b[i] = (float) pel[2];
        pp.setPixel(i, (int) pel[0], (int) pel[1], (int)pel[2], 255);
      }
      return(pp.pels);
  }


}
```

The Mat3 Class

The *Mat3* class is a public class that resides in the *lyon.ipl* package. Its purpose is to provide an optimized way to multiply nonsparse 3x3 matrices. It also acts as a test bed for experiments in using Maple (a symbolic manipulator) to generate optimized Java code.

The code generated by Maple is both human-readable and optimized for speed of execution. When used correctly, the code is also error-free. Maple can ease the programmer's burden in some cases. In general, the code that Maple generates requires human manipulation before it becomes usable as Java.

Class Summary


```
package lyon.ipl;
public class Mat3 {
public Mat3 (double a[][])
  public double [] [] getArray()
  public Mat3 invert()
  public Mat3 multiply(Mat3 bmat3)
  public double[] multiply(double v1, double v2, double v3 )
  public void print()
}
```

Class Usage

There is an example of class usage in the *IYQ* class. The *Mat3* class is designed for multiplying 3 × 3 matrices by each other or by 3 × 1s. It is designed for color-space conversion but is generally applicable to other tasks. The *Mat3* class also supports a fast 3 × 3 matrix inversion. Suppose that the following variables are predefined:

```
double A[][] = {
     { 0.299, 0.587,  0.114},
     { 0.596,-0.274, -0.322},
     { 0.211, 0.522,  0.311}
     };
Mat3  m3;
Mat3 m3inverse;
```

To make a new instance of *Mat3*, use this code:

```
m3 = new Mat3(A);
```

To find the inversion of an instance of *Mat3*:

```
m3inverse = m3.invert();
```

To get the double-precision 3x3 array stored in a *Mat3* instance:

```
A = m3.getArray();
```

To multiply two *Mat3* instances:

```
Mat3 identity = m3.multiply(m3inverse);
```

NOTE

Floating-point error prevents the *Mat3* instance of the identity matrix from being exact, but it is close.

To multiply the 3x3 in *Mat3* by a 3x1 to obtain a 3x1 array of double (also optimized), use this code:

```
double r, g, b;
double v3[] = m3.multiply(r, g, b);
```

To print a *Mat3* instance to *System.out PrintStream*:

```
m3.print();
```

Maple: The Java Gin Joint

Maple is a language for symbolic manipulation [Char et al.]. Maple V was used to generate optimized Java code for implementing matrix manipulation in the *Mat3* class. This is an interesting twist in Java that we have not seen in any other Java books.

The cotton gin is a symbol of the beginning of the industrial revolution. Before the industrial revolution, all work was performed by hand, and assembly-line methods were unknown. Perhaps software is also in its preindustrial-revolution days. We have yet to formulate automatic methods for the generation of software except in special cases. Matrix manipulation is one of those cases (and there are few others).

In Maple, the following code will generate the Java source needed to generate a time-optimal 3x3 matrix inversion:

```
1.      with(linalg):
2.      readlib(C):
3.      a =array(0..2,0..2,[]):
4.      b =array(0..2,0..2,[]):
5.      b =inverse(matrix(a)):
6.      C(b,optimized);
```

The lines are numbered for reference only. Line 1 reads the linear algebra package into Maple. Line 2 reads the C-language generator. Lines 3 and 4 create arrays whose indexes start at zero. Line 5 converts the *a* array into a *matrix* type, forms the inverse symbolically, and sets the answer to *b*. Line 6 outputs the following code:

```
public Mat3 invert() {
   double b[] [] = new double [3][3];
   double t4 = a[0][0]*a[1][1];
   double t6 = a[0][0]*a[1][2];
   double t8 = a[0][1]*a[1][0];
   double t10 = a[0][2]*a[1][0];
   double t12 = a[0][1]*a[2][0];
   double t14 = a[0][2]*a[2][0];
   double t17 =
              1/(-t4*a[2][2]+t6*a[2][1]+t8*a[2][2]
                 -t10*a[2][1]-t12*a[1][2]+t14*a
[1][1]);
   b[0][0] = -(a[1][1]*a[2][2]-a[1][2]*a[2][1])*t17;
   b[0][1] = -(-a[0][1]*a[2][2]+a[0][2]*a[2][1])*t17;
   b[0][2] = (-a[0][1]*a[1][2]+a[0][2]*a[1][1])*t17;
   b[1][0] = (a[1][0]*a[2][2]-a[1][2]*a[2][0])*t17;
   b[1][1] = (-a[0][0]*a[2][2]+t14)*t17;
   b[1][2] = -(-t6+t10)*t17;
   b[2][0] = (-a[1][0]*a[2][1]+a[1][1]*a[2][0])*t17;
   b[2][1] = -(-a[0][0]*a[2][1]+t12)*t17;
   b[2][2] = (-t4-t8)*t17;


   return new Mat3(b);
}
```

We find this code pretty handy. Maple automatically unwinds *for* loops, generates auxiliary variables, and saves a lot of typing. The declaration of *double* for the temporary variables (*tnn*) that Maple generates must be done by hand.

The technique of using Maple extends well into other matrix operations. In Maple, for example, we can unwind the *for* loops in a multiplication:

```
C(multiply(matrix(b),matrix(a)),optimized):
```

The optimized output is reformatted and annotated:

```
public Mat3 multiply(Mat3 bmat3) {
double WW [][] = new double[3][3];
double b [][] = bmat3.getArray();
     WW[0][0] = a[0][0]*b[0][0]+a[0][1]*b[1][0]+a[0][2]*b[2][0];
     WW[0][1] = a[0][0]*b[0][1]+a[0][1]*b[1][1]+a[0][2]*b[2][1];
     WW[0][2] = a[0][0]*b[0][2]+a[0][1]*b[1][2]+a[0][2]*b[2][2];
     WW[1][0] = a[1][0]*b[0][0]+a[1][1]*b[1][0]+a[1][2]*b[2][0];
     WW[1][1] = a[1][0]*b[0][1]+a[1][1]*b[1][1]+a[1][2]*b[2][1];
     WW[1][2] = a[1][0]*b[0][2]+a[1][1]*b[1][2]+a[1][2]*b[2][2];
     WW[2][0] = a[2][0]*b[0][0]+a[2][1]*b[1][0]+a[2][2]*b[2][0];
```

```
    WW[2][1] = a[2][0]*b[0][1]+a[2][1]*b[1][1]+a[2][2]*b[2][1];
    WW[2][2] = a[2][0]*b[0][2]+a[2][1]*b[1][2]+a[2][2]*b[2][2];
    return (new Mat3(WW));
}
```

Generating Java code is not totally automatic. Some human intervention is required—adding the method entry points and returns—but most of the hard work is done by Maple. The following Maple code will generate Java code for any *N*, where *N* is an integer greater than 1:

```
readlib(linalg):
readlib(C):
multn := proc(N)
      local a, b, c, v;
      a := array(0..N,0..N);
      b := array(0..N,0..N);
      c := array(0..N,0..N);
      v := vector([seq(vec[i],i=0..N)]);
      print(v);
      c := multiply(matrix(a),matrix(b));
      print(`Mult N by N times N by N`); print(N+1);
      C(c, optimized);
      print(`Mult N by N times N by 1`); print(N+1);
      c := multiply(matrix(a),v);
      C(c, optimized);
      c := inverse(matrix(a));
      print('inverse'); print(N);
      C(c, optimized);
      end;
```

The time Maple takes to perform the *multn* procedure increases quadratically with increasing *N*, so this technique will not be practical for all sizes of *N*. Also, we have not optimized for especially sparse matrices, such as rotation matrices that occur in computer graphics and robotics.

Image Geometry

Now let's look at the derivation and implementation of elementary 2-D image transformation using matrix multiplication. We will cover the operations of translation, rotation, scale, and shear, using 3x3 matrix multiplications. To perform these operations, we make extensive use of the *Mat3* class derived in the previous section.

Two-Dimensional Translation

To translate a point, *p*, whose coordinates are (*p.x*,*p.y*) plane by an amount, *t*, whose offset is given by, (*t.x*,*t.y*) use the following:

$$p' = (p.x + t.x, p.y + t.y) \tag{9.22}$$

In Java notation:

```
pPrime = p.add(t);
```

An implementation of this translation can be found in the *point* class, which resides in the *lyon* package:

```
public class point extends Computation {
        double x = 0;
        double y = 0;
...

     public point add(point t) {
        return new point(x + t.x, y + t.y);
     }
...
```

This implementation is particularly slow because a new point is allocated every time an addition is performed. We use such code for illustration only.

Two-Dimensional Scaling

To scale a point, *p*, whose coordinates are (*p.x*,*p.y*) plane by an amount, *t*, whose scale is given by, (*s.x*,*s.y*) use the following:

$$p' = (p.x * s.x, p.y * s.y) \tag{9.23}$$

To scale about a point, *t*, translate to the origin, perform the scaling and then translate back. This process can be represented by the following:

$$p' = ((p.x - t.x) * s.x + t.x, (p.x - t.y) * s.y + t.y) \tag{9.24}$$

In Java, we can express Equations 9.24 and 9.23 by overloading a method in the *point* class:

```
public point scale(point s) {
     return new point(x * s.x, y * s.y);
}
```

In the following implementation of Equation 9.24, we scale about point *t*.

```
public point scale(point s, point t) {
    return new point((x - t.x) * s.x + t.x, (y - t.y)* s.y+t.y);
}
```

Because Equation 9.23 scales about the origin, the point will move relative to the origin. Furthermore, if scaling were uniform, the amount of scaling in each dimension would be equal.

Two-Dimensional Rotation

An example of rotation about the center of an image is shown in Figure 9.16.

FIGURE 9.16 LENA ROTATED ABOUT THE CENTER OF THE FRAME.

To rotate about the center of an image, first translate the center to the origin, rotate, and then translate back.

To rotate a point, *p*, whose coordinates are (*p.x*,*p.y*) plane by an amount, θ , about the origin, use the following:

$$p' = (p.x\cos\theta - p.y\sin\theta, p.x\sin\theta + p.y\cos\theta) \tag{9.25}$$

To rotate about a point, *t*, first translate to the origin, perform the rotation, and then translate back. This process can be represented by the following column vector:

$$p' = \begin{bmatrix} (p.x - t.x)(p.x\cos\theta - p.y\sin\theta) + t.x \\ (p.x - t.y)(p.x\sin\theta + p.y\cos\theta) + t.y \end{bmatrix} \tag{9.26}$$

Positive angles are measured counterclockwise. You should be able to reformulate the rotational transformations for negative angles. Use these identities:

$$\cos(-\theta) = \cos\theta$$
$$\sin(-\theta) = -\sin\theta$$

Now let's look at proof of

$$p' = (p.x\cos\theta - p.y\sin\theta, p.x\sin\theta + p.y\cos\theta)$$

Suppose that a complex number of magnitude r represents p so that $p = rei\vartheta$, Then rotation, with respect to the origin, by an amount of θ , can be calculated by multiplication via another complex number, $p\theta = e^i\theta$ so that

$$p' = pp_\theta = re^{i\phi}e^{i\theta} = re^{i(\theta+\phi)} \tag{9.27}$$

By Euler's relation, $e^{i\theta} = \cos\theta + i\sin\theta$, we obtain

$$e^{i(\theta+\phi)} = r[\cos(\theta+\phi) + i\sin(\theta+\phi)] \tag{9.28}$$

We then invoke the double angle formulas for sine and cosine to obtain the following:

$$e^{i(\theta+\phi)} = r\begin{bmatrix}\cos\phi\cos\theta - \sin\phi\sin\theta + \\ i(\cos\phi\sin\theta + \sin\phi\cos\theta)\end{bmatrix} \tag{9.29}$$

Recall that $p = re^{i\phi}$ is, by Euler's relation, the following:

$$p = re^{i\phi} = r\cos\phi + ir\sin\phi = p.x + ip.y \tag{9.30}$$

Substituting Equation 9.30 into Equation 9.29 yields

$$p' = re^{i(\theta+\phi)} = \begin{bmatrix}p.x\cos\theta - p.y\sin\theta + \\ i(p.x\sin\theta + p.y\cos\theta)\end{bmatrix} \tag{9.31}$$

From Equation 9.31, it follows directly that rotation about the origin is given by

$$p' = \left(p.x\cos\theta - p.y\sin\theta, p.x\sin\theta + p.y\cos\theta\right)$$

Q.E.D.

NOTE

A display of the amplitude vs. the phase of a waveform can be rotated by the introduction of a delay.

In the *Shape* class we implement the point rotation about any point using a translation to the center, followed by a rotation, followed by a translation back. The *Shape* class resides in the *lyon* package. The following code modifies the original point so that time does not have to be spent allocating and disposing of new point instances.

```
public void pointRotation(point p, point pc, double theta) {
     // rotate point p about pc an amount of theta radians
     // return the modified point
     double c_theta = Math.cos(theta);
     double s_theta = Math.sin(theta);
     double tx = pc.x + (p.x - pc.x) * c_theta - (p.y - pc.y) * s_theta;
     double ty = pc.y + (p.y - pc.y) * c_theta + (p.x - pc.x) * s_theta;
     p.x = tx;
     p.y = ty;
}
```

The point rotation can be applied to images or graphic objects. Figure 9.17 shows a pinhole camera used for simulating a diffraction range finder in the DiffCAD program. As the camera is repositioned, it automatically pans to point toward the center of a diffraction grating.

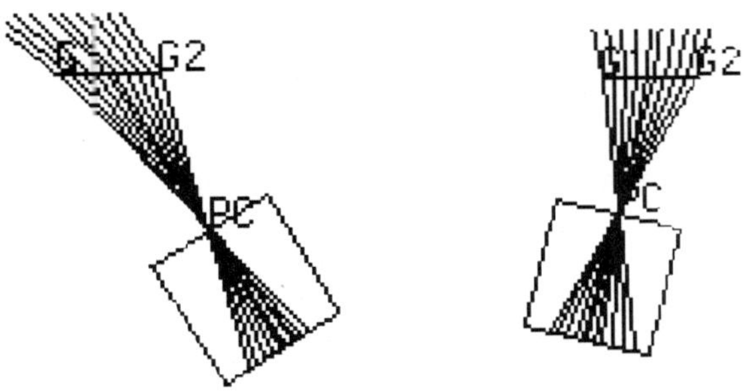

FIGURE 9.17 A PINHOLE CAMERA ROTATED ABOUT ITS CENTER OF FOCUS.

Homogeneous Coordinate Transforms in 2-D

So far we have see that translation, scaling, and rotation are

$$p' = (p.x + t.x, p.y + t.y)$$

$$p' = ((p.x - t.x) * s.x + t.x, (p.x - t.y) * s.y + t.y)$$

and

$$p' = \begin{bmatrix} (p.x - t.x)(p.x\cos\theta - p.y\sin\theta) + t.x \\ (p.x - t.y)(p.x\sin\theta + p.y\cos\theta) + t.y \end{bmatrix}$$

To concatenate several transforms into a single computational entity, we introduce homogeneous coordinates. This approach speeds the computation of combinations of several transformations by creating a single matrix against which points are multiplied. For this discussion, we follow [Foley et al.].

With homogeneous coordinates we use *tuples* in 2-D: (x,y,w). Also, one of the coordinates must be nonzero (typically, w is nonzero).

Iff (if and only if)

$$P = \begin{bmatrix} x \\ y \\ w \end{bmatrix}$$

$$P' = \alpha P \tag{9.32}$$

then P' and P represent the same point. That is, a linear combination of P does not alter the position of the original point. The proportionality factor, α , is eliminated by computing the Cartesian coordinates of the homogeneous point:

$$\text{cartesian coordinates} = (X/W, Y/W, 1) \tag{9.33}$$

The XYW homogeneous coordinate space, with the w=1 plane and point $P(x,y,w)$, projects onto the w=1 plane.

Homogeneous coordinate transformations in 2-D space require a 3x3 matrix multiplication.

NOTE

If all you want to do is to translate, you are doing three multiplications and two additions for nothing!

The matrix form for the translation is as follows:

$$\begin{bmatrix} p'.x \\ p'.y \\ 1 \end{bmatrix} = \begin{bmatrix} 1 & 0 & t.x \\ 0 & 1 & t.y \\ 0 & 0 & 1 \end{bmatrix} \begin{bmatrix} p.x \\ p.y \\ 1 \end{bmatrix} \quad (9.34)$$

This is just like

$$p' = (p.x + t.x, p.y + t.y)$$

with the exception of the w variable.

NOTE

Multiplication by zero in Equation 9.34 takes the computer as long as the multiplication by a nonzero (unless we intercept this special case).

Some graphics textbooks premultiply rather than postmultiply by the column vectors. To convert between the forms, use transposition:

$$\begin{bmatrix} p'.x \\ p'.y \\ 1 \end{bmatrix}^T = \begin{bmatrix} p.x \\ p.y \\ 1 \end{bmatrix}^T \begin{bmatrix} 1 & 0 & t.x \\ 0 & 1 & t.y \\ 0 & 0 & 1 \end{bmatrix}^T \quad (9.35)$$

Expanding Equation 9.35 results in

$$\begin{bmatrix} p'.x & p'.y & 1 \end{bmatrix} = \begin{bmatrix} p.x & p.y & 1 \end{bmatrix} \begin{bmatrix} 1 & 0 & 0 \\ 0 & 1 & 0 \\ t.x & t.y & 1 \end{bmatrix} \quad (9.36)$$

The Java implementation of Equation 9.34 can be found in the translation method in the *Mat3* class of the *lyon.ipl* package:

```
public Mat3 translation(point t) {
     a = new double[3][3];
     a[0][0] = 1;
     a[1][1] = 1;
     a[2][2] = 1;
     a[0][2] = t.x;
     a[1][2] = t.y;
     return new Mat3(a);
}
```

Suppose we write the following:

```
public static void main(String args[]) {
     Mat3 trans = Mat3.translation(new point(2,3));
     trans.print();
}
```

The output is as follows:

```
1 0 2
0 1 3
0 0 1
```

Now we can perform the homogeneous coordinate transformations using matrix multiplications. We use this ability to concatenate many transformations into a single matrix representation. For example:

```
public static void main(String args[]) {
     Mat3 trans1 = Mat3.translation(new point(2,3));
     Mat3 trans2 = Mat3.translation(new point(1,2));
     Mat3 trans3 = trans1.multiply(trans2);
     trans1.print();
     System.out.println(" * " );
     trans2.print();
     System.out.println(" = ");
     trans3.print();
}
1 0 2
0 1 3
0 0 1
 *
1 0 1
0 1 2
0 0 1
 =
1 0 3
0 1 5
0 0 1
```

This code is a concatenation of two transformations—*trans1* and *trans2*—into a single transformation matrix, *trans3*. The matrix product is also called *compounding*, *catenation*, *concatenation*, or *composition*. Once the transform is formulated, it can be applied to all points in the scene.

Several policy issues must be resolved to implement matrix compounding.

We have allocated the memory for a transformation matrix using a static method in the *Mat3* class. The memory allocation is dynamic and possibly wasteful. This might be OK, however, because the number of transformation matrices may be small relative to the number of points that require transformation. When we are finished with the transformation matrix, we should set it to null so that the garbage collector can reclaim the storage.

It is probably good policy to keep a copy of the master object and apply the composite transform to all the points each time. Incrementally transforming the points on an integer coordinate system (such as an image plane) will distort the image. For example, suppose that I wanted to rotate an image by 50(in 10-degree increments. The progression of the image is shown in Figure 9.18.

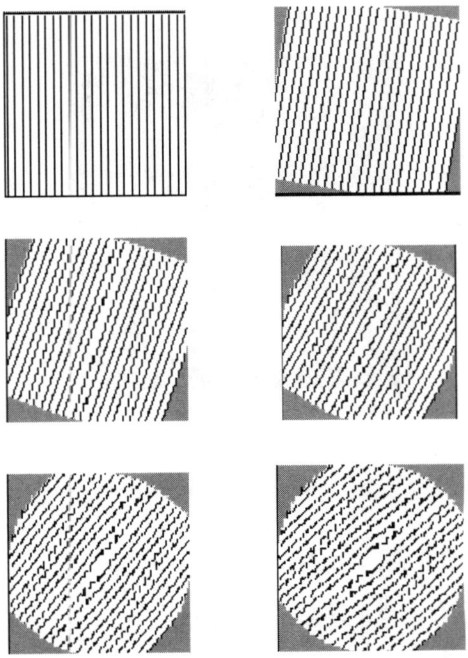

FIGURE 9.18 INCREMENTAL ROTATION IN AN INTEGRAL COORDINATE SYSTEM.

Figure 9.18 shows that when the result of a rotation is used to supply data for the transform input, the result gets distorted and the error accumulates through the transforms. To reduce the distortion, we start with the original data and recompute the transformation matrix. The same 50 degree rotation can be performed, without accumulating distortion, as shown in Figure 9.19.

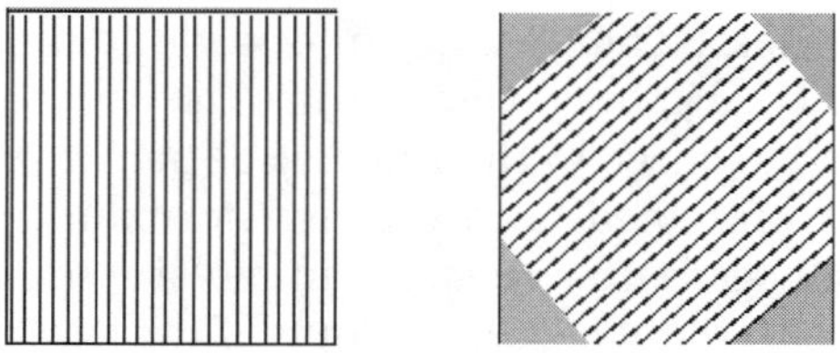

FIGURE 9.19 NON-INCREMENTAL ROTATION IN AN INTEGRAL COORDINATE SYSTEM.

An interesting aspect of the distortion is the aesthetic quality of an image after the incremental rotation. Figure 9.20 shows what happens to the hapless mandrill after a 360 degree rotation is performed in 18 steps of 20 degrees each.

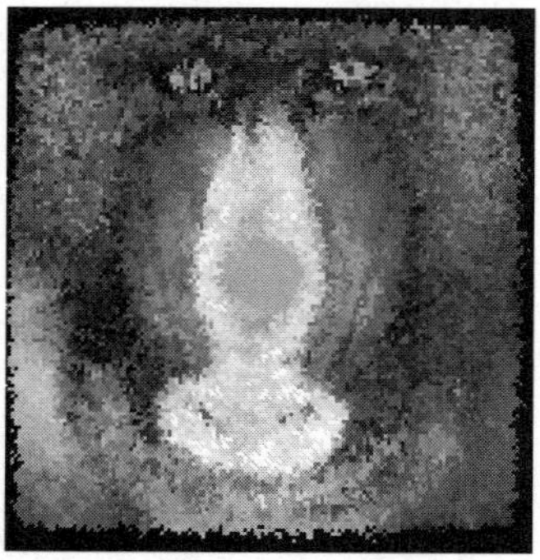

FIGURE 9.20 MANDRILL AFTER 18 INCREMENTAL ROTATIONS OF 20 DEGREES EACH.

The matrix form for the scaling is as follows:

$$\begin{bmatrix} p'.x \\ p'.y \\ 1 \end{bmatrix} = \begin{bmatrix} s.x & 0 & 0 \\ 0 & s.y & 0 \\ 0 & 0 & 1 \end{bmatrix} \begin{bmatrix} p.x \\ p.y \\ 1 \end{bmatrix} \tag{9.37}$$

Here is the Java implementation of Equation 9.37:

```
public static Mat3 scaling(point s) {
     double m[][] = new double[3][3];
     m[0][0] = s.x;
     m[1][1] = s.y;
     m[2][2] = 1;
     return new Mat3(m);
}
```

Like successive translations, successive scalings are multiplicative:

```
public static void main(String args[]) {
     Mat3 trans1 = Mat3.scaling(new point(2,3));
     Mat3 trans2 = Mat3.scaling(new point(1,2));
     Mat3 trans3 = trans1.multiply(trans2);
     trans1.print();
     System.out.println(" * " );
     trans2.print();
     System.out.println(" = ");
     trans3.print();
}
```

Here are the results:

```
2 0 0
0 3 0
0 0 1
 *
1 0 0
0 2 0
0 0 1
 =
2 0 0
0 6 0
0 0 1
```

Figure 9.21 shows the mandrill with a 3:1 zoom out followed by a 3:1 zoom in. This effect was first created optically by systems such as the Blockpix processor [Manning]. Our empirical tests show that the color gamut produced by Blockpix processors has a wider range than that produced by color monitors.

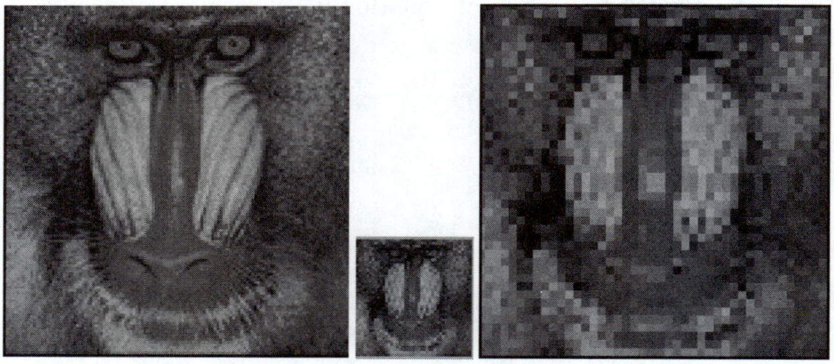

FIGURE 9.21 ORIGINAL MANDRILL (LEFT), ZOOM OUT 3:1, AND ZOOM IN 3:1.

The matrix form for the rotation is as follows:

$$\begin{bmatrix} p'.x \\ p'.y \\ 1 \end{bmatrix} = \begin{bmatrix} \cos\theta & -\sin\theta & 0 \\ \sin\theta & \cos\theta & 0 \\ 0 & 0 & 1 \end{bmatrix} \begin{bmatrix} p.x \\ p.y \\ 1 \end{bmatrix} \tag{9.38}$$

The Java implementation of Equation 9.38 follows:

```
public static Mat3 rotation(double theta) {
     double m[][] = new double[3][3];
     double cas = Math.cos(theta);
     double sas = Math.sin(theta);
     m[0][0] =  cas;
     m[1][1] =  cas;
     m[0][1] = -sas;
     m[1][0] =  sas;
     return new Mat3(m);
}
```

NOTE

The *Math.sin* and *Math.cos* functions in Java use radians for input.

Consider this example:

```
public static void main(String args[]) {
     Mat3 trans3 = Mat3.rotation(90 * piOn180);
     trans3.print();
}
```

The preceding code produces this output:

```
0 -1 0
1 0 0
0 0 0
```

The rotation and translation transforms are rigid-body transformations, because they do not distort the object. Scale is not a rigid-body transformation, because scaling in the x and y directions may be different.

Affine transformations consist of sequences of rotation, translation, scale, and shear operations. Affine transformations preserve the parallelism of lines but not lengths or angles.

Shear can go in the x or y direction. For example, the shear transform in the x direction is given by the following:

(9.39)

The shear transform in the y direction is given by

$$\begin{bmatrix} p'.x \\ p'.y \\ 1 \end{bmatrix} = \begin{bmatrix} 1 & sh.x & 0 \\ 0 & 1 & 0 \\ 0 & 0 & 1 \end{bmatrix} \begin{bmatrix} p.x \\ p.y \\ 1 \end{bmatrix} \tag{9.40}$$

An implementation for the construction of a shear matrix follows:

```
public static Mat3 shear(point sh) {
     double m[][] = new double[3][3];
     m[0][0] = 1;
     m[1][1] = 1;
     m[2][2] = 1;
     m[0][1] = sh.x;
     m[1][0] = sh.y;
     return new Mat3(m);
}
```

The following shows an example of the shear method being invoked:

```
public static void main(String args[]) {
     Mat3 trans1 = Mat3.shear(new point(1,2));
     trans1.print();
}
```

The output follows:

```
1 1 0
2 1 0
0 0 1
```

Applications of Affine Transforms

In this section we show examples of the use of affine transforms in DiffCAD for manipulating image data. In all the cases, the *ProcessPlane* class has been modified to permit the use of a 3x3 matrix transform to process the output coordinates of the image. The output coordinates are multiplied by the matrix transform centered on the image coordinates. The result of the multiplication is used to resample the input image. No filtering is performed, so we can expect aliasing effects. The heart of the transformation is a method called *xform*:

```
public void xform(Mat3 transform) {
 int w = getWidth();
 int h = getHeight();
 int xc = w/2;
 int yc = h/2;
```

The *xform* method takes a precomputed transformation matrix in the form of an instance of the *Mat3* class. The height and width are obtained from the input image. The output image is stored in an instance of *ProcessPlane*, pp:

```
ProcessPlane pp = new ProcessPlane(w,h);
```

The result of the 3x3 matrix multiplication by a 3x1 is a 3x1. The result is stored in an array called *p*:

```
double p[] = new double [3];
int pixel;
int xp, yp;
```

We follow the idea of [Expeset] and start to scan the output image using coordinates that bias the center of the image toward the origin. This technique permits rotation about the center of the image rather than the origin (the upper-left corner of the image in the Java AWT).

```
for (int x = -xc; x < xc; x++)
  for (int y=-yc; y < yc; y++) {
     p=transform.multiply(x,y,1);
     xp = (int) p[0]+xc;
     yp = (int) p[1]+yc;
```

```
      if ((xp < w) && (yp < h) && (xp >= 0) && (yp >= 0)) {
         pixel = getPixel(xp, yp);
         pp.setPixel(x+xc,y+yc,pixel);
      }
  }
  pels = pp.pels;
}
```

The *xform* method permits implementation of any of the transforms by creating a 3x3 matrix and calling *xform*. An example is the *ProcessPlane* method *turn*:

```
public void turn(double degrees) {
 double pion180 = Math.PI / 180.0;
 double theta = degrees * pion180;
 xform(Mat3.rotation(theta));
}
```

Another example is the *ProcessPlane* method *zoom*:

```
public void zoom(double percentage) {
 xform(Mat3.scaling(new point(percentage,percentage)));
}
```

Finally, we implement the shear in *x* and *y* using the following:

```
public void shearx(double shx) {
 xform(Mat3.shear(new point(shx,0)));
}
public void sheary(double shy) {
 xform(Mat3.shear(new point(0,shy)));
}
```

The results of a shear in *x* and a shear in *y* are shown in Figure 9.22.

FIGURE 9.22 SHEAR IN X AND SHEAR IN Y.

Instead of making a new image, we can use the input image as the canvas for our output. We can then reprocess the image using the result. This is known as a *feedback loop*. Many cool effects are based on feedback. Figure 9.23 shows zoom being used with feedback. The only modification is to add a new *xform* method called *xformfeedback* to *ProcessPlane*. The *xformfeedback* method is just like the *xform* method except for one important difference:

```
for (int x = -xc; x < xc; x++)
  for (int y=-yc; y < yc; y++) {
     p=transform.multiply(x,y,1);
     xp = (int) p[0]+xc;
     yp = (int) p[1]+yc;
     if ((xp < w) && (yp < h) && (xp >= 0) && (yp >= 0)) {
        pixel = getPixel(xp, yp);
```

At this point in the code, *xform* would set the pixel on a *ProcessPlane* instance. Instead, *xformfeedback* uses its own.

```
        setPixel(x+xc,y+yc,pixel);
     }
}
```

FIGURE 9.23 MANY COOL EFFECTS ARE BASED ON FEEDBACK.

Originally, analog feedback was obtained by pointing two mirrors at each other. as in a barbershop. The images are scaled in size and placed inside each other. This feedback appears to go on forever. Electronically, the feedback can be performed by pointing a video camera at a monitor that outputs the image scanned by the video camera. This type of processing permits video synthesizers to be placed into the loop. Companies such as Quantel and Ampex manufacture video equipment that can produce such effects in real time [Schwartz et al.]. Figure 9.24 shows Lena after rotational feedback is performed (the process of evolution into the form shown is more interesting than the final form itself). To implement the feedback effect, we invoke the feedback version of the *xform* method:

```
public void turnfb(double degrees) {
 double pion180 = Math.PI / 180.0;
 double theta = degrees * pion180;
 xformfeedback(Mat3.rotation(theta));
}
```

FIGURE 9.24 ROTATIONAL FEEDBACK, 10 ITERATIONS OF 36° EACH.

We can the combine effect sequentially without any modification to DiffCAD. Figure 9.25 shows zoom feedback applied to the result of rotational feedback.

FIGURE 9.25 ROTATIONAL AND ZOOM FEEDBACK (SEQUENTIALLY APPLIED).

With a small modification, we can create a composite transform matrix that can yield composite transform feedback. The following method (*shearyfb*) shows how multiplication of several transform matrices can create a concatenated transform matrix:

```
public void shearyfb(double s) {
     Mat3 t = Mat3.shear(new point(.5,.5));
     point tp = new point(1.4, 0.9);
     t.multiply(Mat3.scaling(tp));
     t.multiply(Mat3.rotation(20*Math.PI/180));
 xformfeedback(t);
}
```

The result of this effect is shown in Figure 9.26.

FIGURE 9.26 COMPOSITE TRANSFORM FEEDBACK.

As a final thought, we propose the theory art is a language of communication and that, like any language, art forms have grammar. For example, camera grammar might be dolly, pan, tilt, and rotate. Thus, we claim that computer art is different from other art forms in that it contributes new effects for image manipulation (such as the affine feedback transform).

Summary

In this chapter we showed how to take a histogram, perform an FFT, perform a IFFT, and use these transforms to filter an image. One application that we explored was the creation of diffraction in the far field. This was made possible by some simplifying assumptions that made far-field diffraction computationally identical to the FFT.

We also introduced an algorithm for raster to vector conversion, an essential first step in extracting linear features from an image. The vectors still require ordering (the Chinese postman problem). Space and time do not permit us to explore a solution, although we have one [Lyon 85].

In this chapter we also explored affine transforms in their typical use: Take an input image, process it, and place it in an output frame. We showed how to make a simple variation on the affine transform by using feedback.

This variation makes for a wonderfully rich aesthetic exploration, often called eye candy. The rotational feedback about the center has a polar-coordinate symmetry that appears to lead to decentralized eye movement. Such patterns appear to have an almost mandala-like effect. We theorize that decentralized eye movement assists the mind in achieving a meditative state. Time did not permit us to develop a nice interface to our image manipulation system, something that might be a topic of future exploration. One aspect of the 2-D image effects that we have not been able to convey in the figures is the aesthetic aspects of the image evolution. During small, incremental steps, intermediate results have been saved to create fascinating experimental animations that permit an exploration of the grammar of affine feedback.

As another whirlwind chapter comes to a close, it is doubly sad, because it marks the end of the book and there is so much left to do! For further exploration of image warping, we refer you to [Wolberg]. For further research into 2-D image effects, [Holtzman] provides a unique perspective.

Image processing is a huge topic, and there are many good books to choose from. We have been looking to [Pratt] and [Myler] for inspiration and guidance.

INDEX

D

E

F

G

H

N

O

P

Q

R

S

T

Colophon

This book was prepared using Microsoft Word 5.1a on a PowerMac 8100/100.

The Java source was transformed into HTML using the DiffCAD HTML generator. The HTML was transformed into Word using MacLink Plus 8.04i from DataViz, Inc. The code was typeset with a monospace typeface.

Equations were set with the Design Science equation editor using MT Extra, Symbol, and Times fonts. The body of the text was set with Times 12.

The class tree layout was done using the Metrowerks class browser. The exception class tree layout was done by hand with assistance from Theo Vosse's TreeParse program, available only for MacOS as far as we know.

Figures were prepared using a screen capture utility called Captivate Select by Main Stay. The screen shots were pasted into Word. The Word document was saved as RTF. Then RTFtoHTML (from http://www.sun-pack.com/RTF) was used to remove the PICT images from the RTF file. Debabelizer was used to convert the color PICT images to the grayscale TIFF images required by the publisher.

The CD-ROM was made using Toast-Pro and an APS Technologies CD-ROM Writer. Toast is probably the best software we have tried.